Essential Statistics for Data Science

Essential Statistics for Data Science

A Concise Crash Course

MU ZHU

Professor, University of Waterloo

OXFORD
UNIVERSITY PRESS

Great Clarendon Street, Oxford, OX2 6DP,
United Kingdom

Oxford University Press is a department of the University of Oxford.
It furthers the University's objective of excellence in research, scholarship,
and education by publishing worldwide. Oxford is a registered trade mark of
Oxford University Press in the UK and in certain other countries

Published in the United States of America by Oxford University Press
198 Madison Avenue, New York, NY 10016, United States of America

British Library Cataloguing in Publication Data

Data available

Library of Congress Control Number: 2023931557

ISBN 978–0–19–286773–5
ISBN 978–0–19–286774–2 (pbk.)

DOI: 10.1093/oso/9780192867735.001.0001

Printed and bound by
CPI Group (UK) Ltd, Croydon, CR0 4YY

To Everest and Mariana

Contents

PART III. FACING UNCERTAINTY

PART IV. APPENDIX

Prologue

When my university first launched a master's program in data science a few years ago, I was given the task of teaching a crash course on statistics for incoming students who have not had much exposure to the subject at the undergraduate level—for example, those who majored in Computer Science or Software Engineering but didn't take any serious course in statistics.

Our Department of Statistics formed a committee, which fiercely debated what materials should go into such a course. (Of course, I should mention that our Department of Computer Science was asked to create a similar crash course for incoming students who did not major in Computer Science, and they had almost exactly the same heated debate.) In the end, a consensus was reached that the statistics crash course should essentially be "five undergraduate courses in one", taught in one semester at a mathematical level that is suitable for master's students in a quantitative discipline.

At most universities, these five undergraduate courses would typically carry the following titles: (i) Probability, (ii) Mathematical Statistics, (iii) Regression, (iv) Sampling, and (v) Experimental Design. This meant that I must somehow think of a way to teach the first two courses—a year-long sequence at most universities—in just about half a semester, and at a respectable mathematical level too. This book is my personal answer to this challenge. (While compressing the other three courses was challenging as well, it was much more straightforward in comparison and I did not have to struggle nearly as much.)

One may ask why we must insist on "a respectable mathematical level". This is because our goal is not merely to teach students some statistics; it is also to warm them up for other graduate-level courses at the same time. Therefore, readers best served by this book are precisely those who not only want to learn some essential statistics very quickly but also would like to continue reading relatively advanced materials that require a decent understanding and appreciation of statistics—including some conference papers in artificial intelligence and machine learning, for instance.

I will now briefly describe some main features of this book. Despite the lightning pace and the introductory nature of the text, a *very deliberate* attempt is still made to ensure that three very important computational techniques—namely, the EM algorithm, the Gibbs sampler, and the bootstrap—are introduced. Traditionally, these topics are almost never introduced to students "immediately" but, for students of data science, there are strong reasons why they should be. If the process of writing this book has been a quest, then it is not an exaggeration for me to say that this particular goal has been its Holy Grail.

To achieve this goal, a great deal of care is taken so as not to overwhelm students with special mathematical "tricks" that are often needed to handle different probability distributions. For example, Part I, *Talking Probability*, uses *only three* distributions—specifically, the Binomial distribution, the uniform distribution on $(0, 1)$, and the normal distribution—to explain all the essential concepts of probability that students will need to know in order to continue with the rest of the book. When introducing multivariate distributions, only their corresponding extensions are used, for example, the multinomial distribution and the multivariate normal distribution.

Then, two much deliberated sets of running examples—specifically, (i) Examples 5.2, 5.4, 5.5, + 5.6 and (ii) Examples 6.2 + 6.3—are crafted in Part II, *Doing Statistics*, which naturally lead students to the EM algorithm and the Gibbs sampler, both in rapid progression and with minimal hustle and bustle. These running examples also use *only two* distributions—in particular, the Poisson distribution, and the Gamma distribution—to "get the job done".

Overall, precedence is given to estimating model parameters in the frequentist approach and finding their posterior distribution in the Bayesian approach, before more intricate statistical questions—such as quantifying how much uncertainty we have about a parameter and testing whether the parameters satisfy a given hypothesis—are then addressed separately in Part III, *Facing Uncertainty*. It's not hard for students to appreciate why we must always try to say something first about the unknown parameters of the probability model—either what their values might be if they are treated as fixed or what their joint distribution might look like if they are treated as random; how else can the model be useful to us otherwise?! Questions about uncertainty and statistical significance, on the other hand, are much more subtle. Not only are these questions relatively uncommon for very complex models such as a deep neural network, whose millions of parameters really have no intrinsic meaning to warrant a significance test, but they also require a unique conceptual infrastructure with its own idiosyncratic jargon (e.g. the p-value).

Finally, some mathematical facts (e.g. Cauchy–Schwarz), stand-alone classic results (e.g. James–Stein), and materials that may be skipped on first reading (e.g. Metropolis–Hastings) are presented through "mathematical inserts", "fun boxes", and end-of-chapter appendices so as to reduce unnecessary disruptions to the flow of main ideas.

PART I
TALKING PROBABILITY

Synopsis: The statistical approach to analyzing data begins with a probability model to describe the data-generating process; that's why, to study statistics, one must first learn to speak the language of probability.

1

Eminence of Models

The very first point to make when we study statistics is to explain why the language of probability is so heavily used.

Everybody knows that statistics is about analyzing data. But we are interested in more than just the data themselves; we are actually interested in the *hidden processes* that produce, or generate, the data because only by understanding the data-generating processes can we start to discover patterns and make predictions. For example, there are data on past presidential elections in the United States, but it is not too useful if we simply go about describing matter-of-factly that only 19.46% of voters in County H voted for Democratic candidates during the past two decades, and so on. It will be much more useful if we can figure out a *generalizable pattern* from these data; for example, people with certain characteristics tend to vote in a certain way. Then, we can use these patterns to predict how people will vote in the next election.

These data-generating processes are described by probability models for many reasons. For example, prior to having seen the data, we have no idea what the data will look like, so the data-generating process appears stochastic from our point of view. Moreover, we anticipate that the data we acquire will inevitably have some variations in them—for example, even people who share many characteristics (age, sex, income, race, profession, hobby, residential neighborhood, and whatnot) will not all vote in exactly the same way—and probability models are well equipped to deal with variations of this sort.

Thus, probability models are chosen to describe the underlying data generating process, and much of statistics is about what we can say about the process itself based on what comes out of it.

At the frontier of statistics, data science, or machine learning, the probability models used to describe the data-generating process can be pretty complex. Most of those which we will encounter in this book will, of course, be much simpler. However, whether the models are complex or simple, this particular characterization of what statistics is about is very important and also why, in order to study statistics at any reasonable depth, it is necessary to become reasonably proficient in the language of probability.

Essential Statistics for Data Science. Mu Zhu, Oxford University Press. © Mu Zhu (2023).
DOI: 10.1093/oso/9780192867735.003.0001

Example 1.1.   Imagine a big crowd of n people. For any two individuals, say, i and j, we know whether they are friends ($x_{ij} = 1$) or not ($x_{ij} = 0$). It is natural to believe that these people would form, or belong to, different communities. For instance, some of them may be friends because they play recreational soccer in the same league, others may be friends because they graduated from the same high school, and so on.

How would we identify these hidden communities? One way to do so would be to postulate that these friendship data, $X = \{x_{ij} : 1 \le i, j \le n\}$, have been generated by a probability model, such as

$$\mathcal{M}(X|Z) = \prod_{i,j} \left(p_{z_i z_j}\right)^{x_{ij}} \left(1 - p_{z_i z_j}\right)^{1-x_{ij}}, \tag{1.1}$$

where $Z = \{z_i : 1 \le i \le n\}$ and each $z_i \in \{1, 2, \ldots, K\}$ is a label indicating which of the K communities individual i belongs to.[1]

Model (1.1) is known as a "stochastic block model" or SBM for short [1, 2, 3]. It states that, independently of other friendships, the probability, $p_{z_i z_j}$, that two individuals i and j will become friends ($x_{ij} = 1$) depends only on their respective community memberships, z_i and z_j. [*Note: The model also states the probability that they will not become friends ($x_{ij} = 0$) is equal to $1 - p_{z_i z_j}$.*]

Table 1.1 shows a hypothetical example. In a small town with its own recreational soccer league and just one local high school, individuals who belong to both communities are highly likely to become friends, with 90% probability. Those who play in the soccer league but didn't go to the local high school are also quite likely to become friends, with 75% probability. But for those who don't belong to either community, the chance of them becoming friends with one another is fairly low, with just 1% probability. And so on.

In reality, only some quantities in the model (i.e. x_{ij}) are *observable*, while others (i.e. z_i, z_j, $p_{z_i z_j}$) are *unobservable*; see Table 1.2. The unobservable quantities must be estimated from the observable ones, and much of statistics is about how this should be done.

In this particular case, a natural way to proceed would be to use an iterative algorithm, alternating between the estimation of $\{z_i : 1 \le i \le n\}$, given $\{p_{k\ell} : 1 \le k, \ell \le K\}$, and vice versa. Once estimated, some of these unobservable

---

[1] Since, at this point, we haven't yet delved into anything formally, including the notion of probability distributions itself, we simply use the *non-standard* notation "$\mathcal{M}$" here to denote the vague idea of "a model". The notation "$X|Z$" actually conforms to the standard conditional notation in probability; here, it simply means that, while both X and Z are quantities that should be described by a probabilistic generating mechanism, this particular model describes only the probability mechanism of X, pretending that Z has been fixed. We will come back to this *type* of model at the end of Chapter 5.

Table 1.1 An illustrative example of the SBM [Model (1.1)]

	$z_j = 1$ (Both)	$z_j = 2$ (Soccer league)	$z_j = 3$ (High school)	$z_j = 4$ (Neither)
$z_i = 1$ (Both)	0.90	0.80	0.20	0.05
$z_i = 2$ (Soccer league)	–	0.75	0.10	0.03
$z_i = 3$ (High school)	–	–	0.30	0.02
$z_i = 4$ (Neither)	–	–	–	0.01

Values of $p_{z_i z_j}$ for $z_i, z_j \in \{1, 2, 3, 4\}$. Symmetric entries (e.g. $p_{21} = p_{12} = 0.80$) are omitted for better visual clarity.

Source: authors.

Table 1.2 Observable and unobservable quantities in the SBM [Model (1.1)] and the ct-SBM [Model (1.2)]

	SBM [Model (1.1)]	ct-SBM [Model (1.2)]
Observables	$\{x_{ij} : 1 \le i, j \le n\}$	$\{t_{ijh} : 1 \le i, j \le n; h = 1, 2, \ldots, m_{ij}\}$
Unobservables	$\{z_i : 1 \le i \le n\}$ $\{p_{k\ell} : 1 \le k, \ell \le K\}^\dagger$	$\{z_i : 1 \le i \le n\}$ $\{p_{k\ell}(t) : 1 \le k, \ell \le K\}^\dagger$

Note: †In fact, the total number of communities, K, is typically also unobservable but, here, we are taking a somewhat simplified perspective by assuming that it is known a priori.

Source: authors.

quantities—specifically, the estimated labels, $\{\hat{z}_i : 1 \le i \le n\}$—can then be used to identify the hidden communities. $\square$

Example 1.2. Now imagine that, instead of knowing explicitly whether people are friends or not, we have a record of when they communicated with (e.g. emailed or telephoned) each other. Table 1.3 shows an example. On November 7, 2016, Amy emailed Bob shortly before midnight, probably to alert him of the imminent election of Donald Trump as the forty-fifth President of the United States; Bob emailed her back early next morning; and so on.

Similarly, one can postulate that these communication data have been generated by a probability model, such as

$$\mathcal{M}(T|Z) = \prod_{i,j} \left\{ \left[e^{-\int_{t_0}^{t_\infty} p_{z_i z_j}(u) du} \right] \times \prod_{h=1}^{m_{ij}} p_{z_i z_j}(t_{ijh}) \right\}, \tag{1.2}$$

Table 1.3 An illustrative example of communication records

From (*i*)		To (*j*)		Time (*t*)
1	(Amy)	2	(Bob)	November 7, 2016, 23:42
2	(Bob)	1	(Amy)	November 8, 2016, 07:11
2	(Bob)	4	(Dan)	November 8, 2016, 07:37
⋮		⋮		⋮

Source: authors.

where $T = \{t_{ijh} : 1 \leq i, j \leq n; h = 1, 2, \ldots, m_{ij}\}$, $t_{ijh} \in (t_0, t_\infty)$ is the time of h-th communication between i and j, and m_{ij} equals the total number of communications between i and j.[2]

Model (1.2) is an extension of (1.1), called a "continuous-time stochastic block model" or ct-SBM for short [4]. It states that, independently of communications between others, the arrival times, $\{t_{ijh} : h = 1, 2, \ldots, m_{ij}\}$, of communications between two individuals i and j are generated by a so-called "non-homogeneous Poisson process" with a rate function, $\rho_{z_i z_j}(t)$, that depends only on their respective community memberships, z_i and z_j. The "non-homogeneous Poisson process"—what appears inside the curly brackets in Equation (1.2)—is a relatively advanced probability model and very much beyond the scope of this book; curious readers who desire a better understanding of this expression can read Appendix 1.A at the end of this chapter at their own risk.

Nevertheless, the parallelism with Example 1.1 is clear enough. Only some quantities (i.e. t_{ijh} now instead of x_{ij}) are *observable*, while others (i.e. z_i, z_j as before but $\rho_{z_i z_j}$ now instead of $p_{z_i z_j}$) are *unobservable*; again, see Table 1.2. The unobservable quantities must be estimated from the observable ones. Here, each $\rho_{z_i z_j}$ is an entire function (of time), so the estimation problem for the ct-SBM [Equation (1.2)] is a lot harder than it is for the "vanilla" SBM [Equation (1.1)], in which each $p_{z_i z_j}$ is simply a scalar.   □

Example 1.3. Interestingly, model (1.2) can be extended and adapted to analyze basketball games. Table 1.4 shows two plays between the Boston Celtics and the Miami Heat that took place in the 2012 NBA Eastern Conference finals, by tracking the movement of the ball from the start of each play to the

---

[2] The notations "t_0" and "t_∞" are used here simply to mean the beginning and end of a time period over which communication patterns are being studied, for example, $t_0 = 00:00$ and $t_\infty = 23:59$.

Table 1.4 Two plays between the Boston Celtics and the Miami Heat that took place in the 2012 NBA Eastern Conference finals

From (i)	To (j)	Time (t, in seconds)
Inbound	C#9	0
C#9	C#5	11
C#5	Miss 2[†]	12
Rebound	H#6	0
H#6	H#3	7
H#3	H#15	8
H#15	H#3	9
H#3	H#6	12
H#6	Miss 3[†]	17

Note: C = Boston Celtics; H = Miami Heat. [†]Missing a two-point (or three-point) shot.

end. It is easily seen that, by and large, these data have the same structure as the communication data displayed in Table 1.3—except that each play necessarily begins from a special state (e.g. `inbound`, `rebound`) and ends in a special state (e.g. `miss 2`, `miss 3`) as well. As such, the model must be extended to deal with these transitions between special states and players.

Another much more subtle, but highly critical, adaptation to the model is also necessary before the model can properly handle basketball games. For example, if LeBron James is not in possession of the ball, the fact that he didn't make a pass at a particular point of time—say, $t_0 \in (0, 24)$ seconds—does *not* contain any information about the underlying rate function at t_0; whereas, the fact that Amy didn't send any email at a certain time t_0—say, midnight—still contains information about the underlying rate function at t_0. The adaptations needed to accommodate subtle differences like these are fairly intricate; for details, see Xin *et al.* [4].

With various extensions and adaptations of this kind, we analyzed two games (Game 1 and Game 5) from the 2012 NBA Eastern Conference finals between the Miami Heat and the Boston Celtics, as well as two games (Game 2 and Game 5) from the 2015 NBA finals between the Cleveland Cavaliers and the Golden State Warriors. For these analyses, we also simplified the rate function $\rho_{k\ell}(t)$ by reparameterizing it to be:

$$\rho_{k\ell}(t) = \lambda_k(t) \times p_{k\ell}$$

for every (k, ℓ) combination, which reduced $K \times K$ functions that must be estimated to just K functions plus $K \times K$ scalars.

By following the movement of the ball in this manner with $K = 3$, we found that both the Boston Celtics and the Miami Heat played with essentially the same three groups of players : (i) point guards; (ii) super stars (Ray Allen and Paul Pierce for the Celtics, Dwyane Wade and LeBron James for the Heat); and (iii) others. However, their respective rate functions, $\lambda_1(t)$, $\lambda_2(t)$, and $\lambda_3(t)$, showed that the two teams played their games very differently (Figure 1.1). For the Miami Heat, the "bump" between $t \in (5, 10)$ in their $\lambda_1(t)$ was because their point guards usually passed the ball to LeBron James and Dwyane Wade and relied on the two of them to organize the offense; whereas, for the Boston Celtics, there was actually a "dip" in their $\lambda_1(t)$ between $t \in (5, 10)$. This was because their point guards—most notably, Rajon Rondo—typically held onto the ball and organized the offense themselves.

In a similar fashion, we found that, during the 2015 finals, the Golden State Warriors played Game 5 very differently from how they had played Game 2. In particular, their $\lambda_3(t)$ changed dramatically between the two games (Figure 1.2). This was almost certainly because the players in this third group had changed as well (Table 1.5). More specifically, two of their players, Andre Iguodala and Draymond Green, appeared to have played very different roles in those two games. Readers who are familiar with the 2014–2015 season of the NBA will be able to recall that, during the 2015 finals, the Golden State Warriors lost both Games 2 and 3, at which point their head coach, Steve Kerr,

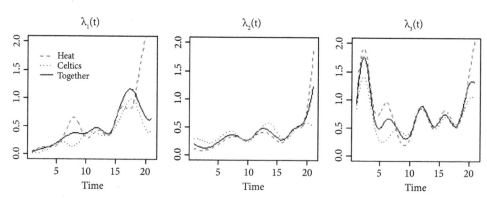

Figure 1.1 Estimated rate functions for three groups of players on the Miami Heat and those on the Boston Celtics, based on Game 1 and Game 5 from the 2012 NBA Eastern Conference finals between the two teams.

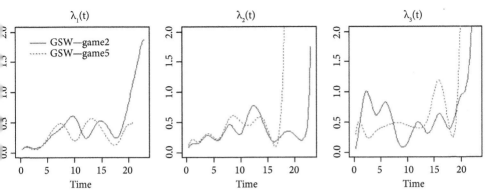

Figure 1.2 Estimated rate functions for three groups of players on the Golden State Warriors, based on Game 2 and Game 5, respectively, from their 2015 NBA finals against the Cleveland Cavaliers.

Source: Reprinted from L. Xin, M. Zhu, H. Chipman (2017). "A continuous-time stochastic block model for basketball networks", *Annals of Applied Statistics*, **11**: 553–597. Copyright 2017, with permission from the Institute of Mathematical Statistics.

Table 1.5 Grouping of players on the Golden State Warriors, based on Game 2 and Game 5, respectively, from their 2015 NBA finals against the Cleveland Cavaliers

Game	$\hat{z}_i = 1$	$\hat{z}_i = 2$	$\hat{z}_i = 3$
#2	PGs Andre Iguodala Draymond Green	SGs	SFs Centers
#5	PGs	SGs SF + PF	Andre Iguodala Draymond Green

PG = point guard (Stephen Curry + Shaun Livingston);
SG = shooting guard (Klay Thompson + Leandro Barbosa);
SF = shooting forward (Harrison Barnes);
PF = power forward (David Lee).

The on-court positions for Iguodala and Green are SF and PF. The analysis groups players by how they *actually* played, rather than how they were *supposed* to have played, each game.

Source: authors.

famously decided to change their regular line-up to a small line-up by stopping to play centers. This was an unconventional strategy, but it successfully turned the series around, and the Warriors went on to win the championship by winning three consecutive games. Our model was apparently capable of detecting this change by simply following the movement of the ball, without explicitly being aware of this piece of information whatsoever.

One reason why we singled out these particular games to analyze was because LeBron James had played both for the 2011–2012 Miami Heat and for the 2014–2015 Cleveland Cavaliers and it was interesting to examine how he played with these two different teams. By creating two separate avatars for him and treating them as two "players" in the model, we analyzed players on these two teams together, using $K = 4$. The four groups of players turned out to be: (i) point guards; (ii) LeBron James of the 2011–2012 Miami Heat, Dwyane Wade, and LeBron James of the 2014–2015 Cleveland Cavaliers; (iii) other perimeter players; and (iv) power forwards and centers. Here, we see that LeBron James is a very special player indeed. With the exception of Dwyane Wade, nobody else on these two teams played like him. By and large, he belonged to a group of his own. In fact, some long-time observers have suggested that his distinctive playing style almost called for the creation of a new on-court position: point forward. Our analysis certainly corroborates such a point of view. □

Through Examples 1.1–1.3 above, we can already get a clear glimpse of the statistical backbone of data science: first, a probability model is postulated to describe the data-generating process; next, the unobservable quantities in the model are estimated from the observable ones; finally, the estimated model is used to reveal patterns, gain insights, and make predictions. It is common for students to think that algorithms are the core of data science, but, in the statistical approach, their role is strictly secondary—they are "merely" incurred by the need to estimate the unobservable quantities in a probability model from the observable ones.

Appendix 1.A For brave eyes only

To better understand the expression inside the curly brackets in Equation (1.2), imagine partitioning the time period (t_0, t_∞) into many tiny intervals such that, on each interval, there is either just one occurrence of the underlying event (here, a communication between i and j) or no occurrence at all. Then, omitting the subscripts, "$z_i z_j$" and "ij", respectively, from $\rho_{z_i z_j}$ and t_{ijh}, the probability that there are occurrences at certain time points and none elsewhere is, in the same spirit as model (1.1), proportional to

$$\prod_{t_h \in \text{no}} [1 - p(t_h)] \times \prod_{t_h \in \text{yes}} p(t_h), \tag{1.3}$$

where the notations "$t_h \in$ yes" and "$t_h \in$ no" mean all those time intervals with and without an occurrence, respectively. But

$$\log \left\{ \prod_{t_h \in \text{no}} [1 - p(t_h)] \right\} = \sum_{t_h \in \text{no}} \log[1 - p(t_h)] \xrightarrow{(\dagger)} \sum_{t_h \in \text{no}} -p(t_h) \xrightarrow{(\ddagger)} \int_{t_0}^{t_\infty} -p(u)du,$$

which is why the first product in Equation (1.3) becomes $[e^{-\int p(u)du}]$ in Equation (1.2). [*Note: The two convergence signs "$\longrightarrow$" above are both results of partitioning the time axis into infinitely many tiny intervals. The first ($\dagger$) is because, on all intervals $t_h \in$ no with no occurrence, the rate function $p(t_h)$ must be relatively small, and $\log(1 - u) \approx -u$ for $u \approx 0$; more specifically, the line tangent to $\log(1 - u)$ at $u = 0$ is $-u$. The second ($\ddagger$) is based on the Riemann sum approximation of integrals.*]

2

Building Vocabulary

This chapter builds the basic vocabulary for speaking the language of probability, from some fundamental laws to the notion of random variables and their distributions. For some students, this will just be a quick review, but it is also possible for those who haven't learned any of these to read this chapter and learn enough in order to continue with the rest of the book.

2.1 Probability

In probability, the *sample space* S is the collection of all possible outcomes when an experiment is conducted, and an *event* $A \subset S$ is a subset of the sample space. Then, the probability of an event A is simply $\mathbb{P}(A) = |A|/|S|$, where "$| \cdot |$" denotes the size of the set.

Example 2.1. For example, if we independently toss two regular, unloaded, six-faced dice, the sample space is simply

$$S = \{ \begin{array}{cccccc} (1,1), & (1,2), & (1,3), & (1,4), & (1,5), & (1,6), \\ (2,1), & (2,2), & (2,3), & (2,4), & (2,5), & (2,6), \\ (3,1), & (3,2), & (3,3), & (3,4), & (3,5), & (3,6), \\ (4,1), & (4,2), & (4,3), & (4,4), & (4,5), & (4,6), \\ (5,1), & (5,2), & (5,3), & (5,4), & (5,5), & (5,6), \\ (6,1), & (6,2), & (6,3), & (6,4), & (6,5), & (6,6) \end{array} \},$$

a collection of all possible outcomes, and the event $A = \{$obtain a sum of 10$\}$ is simply the subset

$$A = \{(4,6),(5,5),(6,4)\} \subset S,$$

so $\mathbb{P}(A) = 3/36$.

Likewise, the event $B = \{$the two dice do not show identical result$\}$ is simply the subset

$$B = S \backslash \{(1,1),(2,2),\ldots,(6,6)\},$$

Essential Statistics for Data Science. Mu Zhu, Oxford University Press. © Mu Zhu (2023).
DOI: 10.1093/oso/9780192867735.003.0002

where "$A \backslash B$" denotes set subtraction, that is, $A \backslash B \equiv A \cap B^c$, so $\mathbb{P}(B) = (36 - 6)$ /36 = 30/36.   □

This basic notion of probability here explains why the study of probability almost always starts with some elements of combinatorics, that is, how to count. Indeed, counting correctly can be quite tricky sometimes, but it is not a topic that we will cover much in this book. Some rudimentary experience with basic combinatorics, say, at a high-school level, will be more than enough to read this book.

2.1.1 Basic rules

We state some "obvious" rules. For A, $B \subset S$,

(a) $\mathbb{P}(S) = 1, \mathbb{P}(\phi) = 0, 0 \leq \mathbb{P}(A) \leq 1$, where ϕ denotes the empty set;
(b) $\mathbb{P}(A^c) = 1 - \mathbb{P}(A)$, where A^c denotes the complement of A or the event {not A};
(c) $\mathbb{P}(A \cup B) = \mathbb{P}(A) + \mathbb{P}(B) - \mathbb{P}(A \cap B)$, where $A \cup B$ denotes the event {A or B} and $A \cap B$ the event {A and B}.

Rules (a)–(b) above hardly require any explanation, whereas rule (c) can be seen by simply drawing a *Venn diagram* (Figure 2.1).

Exercise 2.1. Use the Venn diagram to convince yourself of the following identities:

$$A^c \cap B^c = (A \cup B)^c \quad \text{and} \quad A^c \cup B^c = (A \cap B)^c.$$

These are known as *De Morgan's laws*.   □

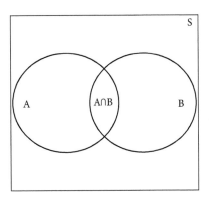

Figure 2.1 A Venn diagram.
Source: authors.

2.2 Conditional probability

A very important concept is the notion of conditional probability.

Definition 1 (Conditional probability). *The quantity*

$$\mathbb{P}(A|B) = \frac{\mathbb{P}(A \cap B)}{\mathbb{P}(B)} \tag{2.1}$$

is called the conditional probability of A given B.   □

It is useful to develop a strong intuitive feel for why the conditional probability is so defined. If we know B has occurred, then this additional piece of information effectively changes our sample space to B because anything outside the set B is now irrelevant. Within this new, effective sample space, only a subset of A still remains—specifically, the part also shared by B, that is, $A \cap B$. Thus, the effect of knowing "B has occurred" is to restrict the sample space S to B and the set A to $A \cap B$.

Example 2.2. Recall Example 2.1 in section 2.1. What happens to the probability of A if we know B has occurred?[1] If B has occurred, it effectively changes our sample space to

$$\begin{array}{cccccc}
\{ & - & (1,2), & (1,3), & (1,4), & (1,5), & (1,6), \\
& (2,1), & - & (2,3), & (2,4), & (2,5), & (2,6), \\
& (3,1), & (3,2), & - & (3,4), & (3,5), & (3,6), \\
& (4,1), & (4,2), & (4,3), & - & (4,5), & (4,6), \\
& (5,1), & (5,2), & (5,3), & (5,4), & - & (5,6), \\
& (6,1), & (6,2), & (6,3), & (6,4), & (6,5), & - & \}
\end{array}$$

because the elements $(1,1), (2,2), \ldots, (6,6)$ are now impossible. In this new, effective sample space (of size 30), how many ways are there for A (a sum of 10) to occur? Clearly, the answer is two—$(4,6)$ and $(6,4)$. So the conditional probability of A, given B, is $\mathbb{P}(A|B) = 2/30$.   □

2.2.1 Independence

Intuitively, it makes sense to say that two events (e.g. A and B) are *independent* if knowing one has occurred turns out to have no impact on how likely the

---

[1] This example has been adapted from the classic text by Sheldon Ross [5].

other will occur, that is, if

$$\mathbb{P}(A|B) = \mathbb{P}(A). \tag{2.2}$$

By Equation (2.1), this is the same as

$$\mathbb{P}(A \cap B) = \mathbb{P}(A)\mathbb{P}(B). \tag{2.3}$$

That's why we are often told that "independence means factorization".
   A trivial rearrangement of Equation (2.1) gives

$$\mathbb{P}(A \cap B) \quad = \quad \mathbb{P}(A|B)\mathbb{P}(B)$$
$$\text{or} \quad \mathbb{P}(B|A)\mathbb{P}(A). \tag{2.4}$$

These are actually how *joint probabilities* (of A and B) must be computed in general when independence (between A and B) cannot be assumed.

Exercise 2.2.   Ellen and Frank have a meeting. Let

$$E = \{\text{Ellen is late}\} \quad \text{and} \quad F = \{\text{Frank is late}\}.$$

Suppose $\mathbb{P}(E) = 0.1$ and $\mathbb{P}(F) = 0.3$. What is the probability that they can meet on time:

(a) if E is independent of F;
(b) if $\mathbb{P}(F|E) = 0.5 > \mathbb{P}(F)$;
(c) if $\mathbb{P}(F|E) = 0.1 < \mathbb{P}(F)$?

In which case—(a), (b), or (c)—is the probability (of them meeting on time) the highest? Does this make intuitive sense?   ☐

Remark 2.1.   This toy problem nevertheless illustrates something much deeper. Often, there are multiple risk factors affecting the probability of a desired outcome, and the answer can be very different whether these risk factors are operating (i) independently of each other, (ii) in the "same direction", or (iii) in "opposite directions". To a large extent, misjudging how different risk factors affected each other was why the 2008 financial crisis shocked many seasoned investors.   ☐

2.2.2 Law of total probability

If the sample space S is partitioned into disjoint pieces $B_1, B_2, \ldots, B_n$ such that

$$S = \bigcup_{i=1}^{n} B_i \quad \text{and} \quad B_i \cap B_j = \phi \quad \text{for} \quad i \neq j,$$

then, as can be easily seen from Figure 2.2,

$$\mathbb{P}(A) = \sum_{i=1}^{n} \mathbb{P}(A \cap B_i) \overset{\star}{=} \sum_{i=1}^{n} \mathbb{P}(A|B_i)\mathbb{P}(B_i), \tag{2.5}$$

where the step marked by "$\star$" is due to Equation (2.4). Equation (2.5) is known as the *law of total probability*.

Even though this law is pretty easy to derive, its implications are quite profound. It gives us a very powerful strategy for computing probabilities—namely, if the probability of something (e.g. A) is hard to compute, look for extra information (e.g. $B_1, B_2, \ldots, B_n$) so that the conditional probabilities, given various extra information, $\mathbb{P}(A|B_1), \mathbb{P}(A|B_2), \ldots, \mathbb{P}(A|B_n)$, may be easier to compute; then, piece everything together.

While all of this may still sound rather straightforward, true mastery of this technique is not easy to acquire without considerable experience, as the following example will demonstrate.

Example 2.3.   A deck of randomly shuffled cards contains n "regular" cards plus one joker. (For convenience, we will refer to such a deck as D_n.) You and I take turns to draw from this deck (without replacement). The one who draws

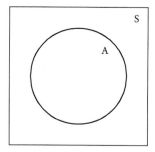 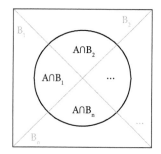

Figure 2.2   Illustrating the law of total probability.
Source: authors.

the joker first will win a cash prize. You go first. What's the probability you will win?

At first glance, the situation here seems rather complex. On second thoughts, we realize that the situation is easier when n is relatively small. In fact, the extreme case of $n = 1$ is trivial. The answer is simply 1/2 if you draw from D_1—with equal chance, either you draw the joker and win immediately or you draw the only other card, in which case, I will draw the joker next and you are sure to lose.

What happens when you draw from D_2 instead? Let

$$W = \{\text{you win}\}.$$

What extra information can we look for to help us pinpoint the probability of W? What about the outcome of your very first draw? Let

$$J = \{\text{your first draw is the joker}\}.$$

Then, by the law of total probability,

$$\mathbb{P}(W) = \mathbb{P}(W|J)\mathbb{P}(J) + \mathbb{P}(W|J^c)\mathbb{P}(J^c). \tag{2.6}$$

Clearly, if your first draw is already the joker, then you win immediately; so $\mathbb{P}(W|J) = 1$. If your first draw is not the joker, then it's my turn to draw. But I now draw from a reduced deck, D_1, since you have already drawn a "regular" card. As we have already argued above, my chance of winning while drawing from D_1 is 1/2, so your chance of winning at this point is $1 - 1/2 = 1/2$. Thus, Equation (2.6) becomes

$$\mathbb{P}(W) = (1)\left(\frac{1}{3}\right) + \left(1 - \frac{1}{2}\right)\left(\frac{2}{3}\right) = \frac{2}{3}.$$

It is not hard to see now that this reasoning process can be carried forward inductively—if you don't draw the joker immediately from D_n, then it's my turn to draw from D_{n-1}. Let

$$p_n = \mathbb{P}(\text{you win with } D_n).$$

Then, the inductive step is

$$p_n = (1)\left(\frac{1}{n+1}\right) + (1 - p_{n-1})\left(\frac{n}{n+1}\right),$$

with $p_1 = 1/2$ being the baseline case. $\qquad\qquad\qquad\qquad\qquad\square$

Exercise 2.3. Continue with Example 2.3 and show that

$$
p_n = \begin{cases} \dfrac{1}{2}, & \text{if } n \text{ is odd}; \\[2ex] \dfrac{n+2}{2n+2}, & \text{if } n \text{ is even} \end{cases}
$$

by completing the inductive argument. □

2.2.3 Bayes law

Beginning students are sometimes tempted to think that $\mathbb{P}(A|B) = \mathbb{P}(B|A)$. It is important to keep in mind that this is definitely *not* the case! Suppose $A = \{\text{rain}\}$ and $B = \{\text{see dark clouds}\}$. If you see dark clouds, there is a chance that it might soon rain, but it might not—it could just remain overcast for the whole day. However, if it rains, then you almost always see dark clouds. Hence, $\mathbb{P}(A|B) < \mathbb{P}(B|A)$.

What is the relationship between these two conditional probabilities in general? By Definition 1 and Equation (2.4),

$$
\mathbb{P}(B|A) = \frac{\mathbb{P}(B \text{ and } A)}{\mathbb{P}(A)} = \frac{\mathbb{P}(A|B)\mathbb{P}(B)}{\mathbb{P}(A)}. \tag{2.7}
$$

Equation (2.7) makes the relationship between the two conditional probabilities explicit.

If $B_1, B_2, \ldots, B_n$ form a partition of the sample space S (see section 2.2.2), then by Equation (2.7) above and the law of total probability, Equation (2.5), we have:

$$
\mathbb{P}(B_i|A) = \frac{\mathbb{P}(A|B_i)\mathbb{P}(B_i)}{\mathbb{P}(A)} = \frac{\mathbb{P}(A|B_i)\mathbb{P}(B_i)}{\sum_j \mathbb{P}(A|B_j)\mathbb{P}(B_j)}. \tag{2.8}
$$

Equations (2.7) and (2.8) are known as *Bayes law*. Like the law of total probability, Bayes law is almost trivial to derive as well, but its implications are equally profound.

Example 2.4. A fraud-detection algorithm flags a transaction as positive if it deems it to be fraudulent. Suppose that the overall *prevalence* of frauds is 0.5%, and the algorithm has a *true positive rate* of 98% and a *false positive rate* of 1%. That is,

$$\mathbb{P}(F) = 0.5\%, \quad \mathbb{P}(\oplus|F) = 98\%, \quad \text{and} \quad \mathbb{P}(\oplus|F^c) = 1\%,$$

where

$$F = \{\text{transaction is fraudulent}\}, \quad F^c = \{\text{transaction is } not \text{ fraudulent}\},$$

and

$$\oplus = \{\text{algorithm flags the transaction as positive}\}.$$

Surely an algorithm with a true positive rate of close to 100% and a false positive rate of close to 0% must be a pretty good algorithm. But if the algorithm has just flagged a transaction as positive, what's the probability that it actually is fraudulent?

By Bayes law, we see that

$$
\begin{aligned}
\mathbb{P}(F|\oplus) &= \frac{\mathbb{P}(\oplus|F)\mathbb{P}(F)}{\mathbb{P}(\oplus|F)\mathbb{P}(F) + \mathbb{P}(\oplus|F^c)\mathbb{P}(F^c)} \\
&= \frac{(0.98)(0.005)}{(0.98)(0.005) + (0.01)(1 - 0.005)} \\
&\approx 33\%.
\end{aligned}
$$

That is, even if such a good algorithm flags a transaction as positive, there is still only about a 33% chance that it actually is fraudulent!

Of course, this does *not* mean that the algorithm is useless. For a randomly selected transaction, there is only a 0.5% chance of being fraudulent. But if it is flagged by the algorithm, it now has a 33% chance of being so—that's a 66-fold increase!

One can actually learn a very important practical lesson from this example. There is a difference between *relative risk* and *absolute risk*. Relative to any transaction, a transaction that has been flagged by the algorithm has a much higher risk of being a fraud; however, this alone does not automatically imply that the absolute risk must also be high.[2] □

---

[2] So, you love a certain unhealthy (but delicious) food. Eating it regularly may significantly increase your risk of developing a certain disease, say, a 10-fold increase from 0.1% to 1%. Should you stop eating the food because of the 10-fold increase in health risk, or should you continue to savor it because your elevated risk is still merely 1%? Only you can make this decision for yourself!

Exercise 2.4. A certain species of rat is either black or brown. If a rat is black, its genotype could be either BB or Bb with equal probability. If a rat is brown, its genotype must be bb. Suppose a male black rat mates with a female brown rat.

(a) If they have one offspring and it is black, what is the probability that the father has genotype BB?
(b) If they have two offspring and they are both black, what is the probability that the father has genotype BB?
(c) Why does it make intuitive sense that the probability in part (b) should be smaller/larger than the one in part (a)?

[*Hint: For those not familiar with basic genetics, it suffices to know that, if Bb mates with bb, their offspring can be Bb or bb with equal probability; whereas, if BB mates with bb, their offspring can only be Bb.*] □

2.3 Random variables

As the name suggests, a *random variable* is a variable whose value is random; it can take on different values with different probabilities.[3] Therefore, each random variable must have a *probability distribution* associated with it so that its behavior can be characterized and described. For example, even though we don't know what the value will turn out to be, from its distribution we may know that there is a higher chance for the value to be between 8.7 and 9.1 than for to it to be between 0.4 and 1.0, and so on.

A *discrete* random variable X takes values on a *countable* (but possibly infinite) set, for example, $\{0.1, 0.2, 0.3\}$ or $\{0, 1, 2, \ldots\}$. Associated with it is a *probability mass function,*

$$f(x) = \mathbb{P}(X = x), \tag{2.9}$$

to describe the probabilities of it taking on different possible values.

---

[3] Mathematically, a random variable is defined as a function $X : S \to \mathbb{R}$ that maps the sample space to the real line. For example, if we toss two coins, then $S = \{HH, HT, TH, TT\}$, and a random variable X such that $X(HH) = 2$, $X(HT) = X(TH) = 1$, and $X(TT) = 0$ simply counts the number of heads we obtain in total. Its value is random because the outcome of the coin toss itself is random. This strict mathematical definition becomes crucial when one studies the so-called "almost sure convergence" of random variables, but we will not be going into that kind of depth in this book.

By contrast, a *continuous* random variable X takes values on an *uncountable* set. For example, $(0, 1)$, $[0, \infty)$ or the entire real line $\mathbb{R}$. Associated with it is a *probability density function*,

$$f(x) = \lim_{\Delta x \downarrow 0} \frac{\mathbb{P}(x \leq X < x + \Delta x)}{\Delta x}, \tag{2.10}$$

to describe the relative likelihood of it being near any given point x. It is clear from Equation (2.10) above that, for a continuous random variable,

$$\mathbb{P}(X = x) = \lim_{\Delta x \downarrow 0} \mathbb{P}(x \leq X < x + \Delta x) = \lim_{\Delta x \downarrow 0} f(x)\Delta x = 0.$$

That is, for a continuous random variable X, the probability of it being exactly equal to any fixed value x is simply zero.

2.3.1 Summation and integration

Mathematicians find it unsatisfying that different types of functions—mass functions and density functions—must be used to describe the distributions of discrete and continuous random variables. It is possible to describe both types of random variables with just one type of function, defined below.

Definition 2 (Cumulative distribution function). *For any random variable X, the function*

$$F(x) = \mathbb{P}(X \leq x) \tag{2.11}$$

is called its cumulative distribution function (CDF). □

In this book, we will always prefer intuitive appeal to mathematical elegance—although the two ideals are not always at odds with each other; often what is mathematically elegant is also intuitively appealing. For discrete random variables, the CDF is not too useful in practice, and it is much more intuitive to think directly in terms of their mass functions.

For continuous random variables, however, the CDF is useful because density functions do not give us probabilities directly, whereas CDFs do. (Mass functions also give us probabilities directly, which explains why CDFs are much less important for discrete random variables.)

In particular, the probability appearing on the right-hand side of Equation (2.10) can be computed using the CDF, while noting that $\mathbb{P}(X = x) = \mathbb{P}(X = x + \Delta x) = 0$:

$$\mathbb{P}(x \leq X < x + \Delta x) = F(x + \Delta x) - F(x).$$

Hence, Equation (2.10) implies a fundamental relationship between the density function and the CDF for a continuous random variable:

$$f(x) = \frac{d}{dx}F(x) \quad \Rightarrow \quad F(x) = \int_{-\infty}^{x} f(t)\, dt. \tag{2.12}$$

Thus, for a continuous random variable, to compute the probability that it is between a and b, we simply *integrate* its density function from a to b:

$$\mathbb{P}(a \leq X \leq b) = F(b) - F(a) = \int_{a}^{b} f(x)\, dx.$$

For a discrete random variable, this is a lot more intuitive—we simply add up all the probabilities between a and b; that is, we *sum* its mass function from a to b:

$$\mathbb{P}(a \leq X \leq b) = \sum_{x=a}^{b} f(x).$$

This parallelism between *integration* (for continuous random variables) and *summation* (for discrete random variables) is very fundamental for the study of probability. We should simply think of them as being the same operation. For example, it is clear why properly defined mass functions and density functions must satisfy:

$$\sum_{x} f(x) = 1 \quad \text{and} \quad \int f(x)\, dx = 1,$$

respectively. Personally, I prefer to use the word "aggregate" to mean both "integrate" and "sum", depending on whether the underlying random variable is continuous or discrete.

2.3.2 Expectations and variances

We briefly state two important summaries of a probability distribution.

Definition 3 (Expectation). *The expectation (or the mean) of a random variable X is defined as*

$$\mathbb{E}(X) \;=\; \sum_x x f(x) \quad or \quad \int x f(x)\, dx, \qquad (2.13)$$

depending on whether it is discrete or continuous. □

Definition 4 (Variance). *For any random variable X, the quantity*

$$\mathbb{V}ar(X) \;=\; \mathbb{E}\{[X - \mathbb{E}(X)]^2\} \qquad (2.14)$$

is called its variance. □

The expectation $\mathbb{E}(X)$ describes the center of the distribution; it tells us what the value of X is *on average*. The variance $\mathbb{V}ar(X)$ is defined directly as the average squared distance between X and the center of its distribution, $\mathbb{E}(X)$; it is, thus, a measure of how *spread out* the distribution is.

Remark 2.2. Sometimes, the quantity $\sqrt{\mathbb{V}ar(X)}$—called the *standard deviation of X*—is used instead of $\mathbb{V}ar(X)$ itself. [*Think: Why is this sometimes more desirable?*] □

The following result is easy to prove, so we will leave it as an exercise.

Exercise 2.5. Show that, if a, b are (non-random) constants, then $\mathbb{E}(aX + b) = a\mathbb{E}(X) + b$ and $\mathbb{V}ar(aX + b) = a^2\mathbb{V}ar(X)$. □

Using the results from Exercice 2.5 above, and noticing that $\mathbb{E}(X)$ is a non-random quantity, we can derive that

$$\mathbb{V}ar(X) = \mathbb{E}\{[X - \mathbb{E}(X)]^2\} = \mathbb{E}\{X^2 - 2X\mathbb{E}(X) + [\mathbb{E}(X)]^2\}$$
$$= \mathbb{E}(X^2) - 2\mathbb{E}(X)\mathbb{E}(X) + [\mathbb{E}(X)]^2 = \mathbb{E}(X^2) - [\mathbb{E}(X)]^2. \qquad (2.15)$$

Equation (2.15) is often used in practice to compute the variance rather than Equation (2.14) itself.

Exercise 2.6 below shows that, knowing just its expectation $\mathbb{E}(X)$ and variance $\mathbb{V}\mathrm{ar}(X)$, but nothing else, we can already say a fair bit about the entire distribution of X—namely, the chances for a random variable to be very far (measured in standard-deviation units) from its center cannot be very large.

Exercise 2.6. Suppose $\mu \equiv \mathbb{E}(X) < \infty$ and $\sigma^2 \equiv \mathbb{V}\mathrm{ar}(X) > 0$. Prove *Chebyshev's inequality*, which states

$$\mathbb{P}(|X - \mu| \geq k\sigma) \leq \frac{1}{k^2}$$

for any $k > 0$. (Of course, the result is truly meaningful only for $k > 1$.) [*Hint: Start by computing the variance of X as*

$$\sigma^2 = \mathbb{V}\mathrm{ar}(X) = \int (x - \mu)^2 f(x)dx = \int_{-\infty}^{\mu-k\sigma} (x - \mu)^2 f(x)dx$$

$$+ \int_{\mu-k\sigma}^{\mu+k\sigma} (x - \mu)^2 f(x)dx + \int_{\mu+k\sigma}^{+\infty} (x - \mu)^2 f(x)dx.$$

Then, reduce the right-hand side to "create" the inequality.]   □

2.3.3 Two simple distributions

Let us look at two specific distributions, one for a discrete random variable and another for a continuous one.

Example 2.5. A random variable X is said to follow the Binomial(n, p) distribution if it has probability mass function

$$f(x) = \mathbb{P}(X = x) = \binom{n}{x}p^x(1 - p)^{n-x}, \quad x \in \{0, 1, 2, \ldots, n\}. \quad (2.16)$$

This is the model that describes n independent coin tosses, and X is the total number of heads obtained (see Figure 2.3). Each toss has probability p of turning up heads and probability $1 - p$ of turning up tails. If we have obtained a total of x number of heads, the remaining $n - x$ tosses must all be tails. Since the tosses are all independent, this explains the term $p^x(1-p)^{n-x}$ in Equation (2.16) above.

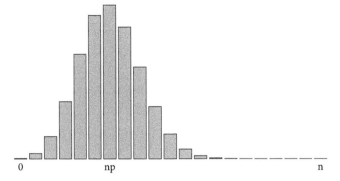

Figure 2.3 The probability mass function for
Binomial(n, p).

Note: In this illustration, np "happens to be" an integer itself, but it doesn't
have to be in general, in which case the peak of the mass function will not
exactly be at np but simply "nearby".
Source: authors.

But there is more than one way to obtain a total of x number of heads. For
example, if $n = 4$ and $x = 2$, then the two heads could have happened on the
first and second tosses, on the first and third tosses, on the first and fourth
tosses, on the second and third tosses, on the second and fourth tosses, or on
the third and fourth tosses. In general, there are $\binom{n}{x} = n!/[x!(n-x)!]$ ways for
us to obtain x number of heads, and, hence, the term $\binom{n}{x}$, in Equation (2.16).

For simplicity, let us now consider the special case of $n = 1$. (The more
general case is considered in Exercise 3.11 later.) Thus, X is now the result of
just *one* coin toss. Then,

$$\mathbb{E}(X) = \sum_{x=0}^{1} x \cdot p^x (1-p)^{1-x} = (0)(1-p) + (1)(p) = p$$

and

$$\mathbb{V}\text{ar}(X) = \mathbb{E}[(X-p)^2] = \sum_{x=0}^{1} (x-p)^2 \cdot p^x (1-p)^{1-x}$$

$$= (0-p)^2(1-p) + (1-p)^2(p) = p(1-p).$$

Notice that, here, $\mathbb{V}\text{ar}(X) = 0$ when $p = 0$ or $p = 1$ and reaches its maximum
when $p = 1/2$. [*Think: Why?*] This shows us, intuitively, why the variance
can be regarded as a measure of *uncertainty*. When $p = 0$ or 1, there is no
uncertainty and we can predict the outcome of the coin flip perfectly—either
tails or heads, but for sure. If $p = 0.9$ (or 0.1), even though the outcome is no
longer certain, we are still assured that one outcome is a lot more likely than

the other. The amount of uncertainty is maximal when $p = 1/2$, when both outcomes are equally likely, and it becomes truly impossible for us to predict ahead of time which outcome will occur. □

Remark 2.3. In the special case of $n = 1$, the Binomial$(1, p)$ distribution is also referred to as the Bernoulli(p) distribution. □

Example 2.6. A random variable X is said to follow the Uniform$(0, 1)$ distribution if it has probability density function

$$f(x) = 1, \quad x \in (0, 1).$$

The density function being flat merely says that all values on $(0, 1)$ are *equally likely*, which is where the name "uniform" comes from. It is easy to compute that

$$\mathbb{E}(X) = \int_0^1 x f(x)\, dx = \int_0^1 x\, dx = \left[\frac{x^2}{2}\right]_0^1 = \frac{1}{2}$$

is simply the middle point between 0 and 1. [*Think: Does this make sense?*] To compute its variance, we first compute

$$\mathbb{E}(X^2) = \int_0^1 x^2 f(x)\, dx = \int_0^1 x^2\, dx = \left[\frac{x^3}{3}\right]_0^1 = \frac{1}{3},$$

from which we can conclude that

$$\mathbb{V}\mathrm{ar}(X) = \mathbb{E}(X^2) - [\mathbb{E}(X)]^2 = \frac{1}{3} - \left[\frac{1}{2}\right]^2 = \frac{1}{12},$$

using Equation (2.15). □

Remark 2.4. In Example 2.5, it is implicit that the probability mass function $f(x) = 0$ for all $x \notin \{0, 1, 2, \ldots, n\}$. Likewise, in Example 2.6, it is implicit that the probability density function $f(x) = 0$ for all $x \notin (0, 1)$. Since every $f(x)$ must be non-negative, it suffices to specify only where it is positive; everywhere else, one can assume it is simply zero. □

Exercise 2.7. Suppose $X \sim$ Uniform(a, b); that is, X is a continuous random variable equally likely to take on any value between a and b. Find its density function $f(x)$, expectation $\mathbb{E}(X)$, and variance $\mathbb{V}\text{ar}(X)$. [*Think: Do the answers for* $\mathbb{E}(X)$ *and* $\mathbb{V}\text{ar}(X)$ *make intuitive sense?*]   □

Exercise 2.8. Suppose $X \sim$ Uniform$\{1, 2, \ldots, n\}$; that is, X is a discrete random variable equally likely to take on any value in the set $\{1, 2, \ldots, n\}$. Find its mass function $f(x)$, expectation $\mathbb{E}(X)$, and variance $\mathbb{V}\text{ar}(X)$.   □

2.4 The bell curve

It is impossible to study probability and statistics without knowing at least something about the famous "bell curve" distribution—officially known as the *normal distribution*. Because it is used literally "everywhere", there are many "standard tricks" for handling such a distribution. Even though it is absolutely essential to master at least some of these tricks if one wants to become a serious student of data science, in this book we take the viewpoint that it is better to first focus on the main principles and ideas and not be distracted by particular mathematical tricks, however important they may be. That is why our treatment of the normal distribution actually will be much lighter than is typical. In fact, we will simply state the most important facts about it in the box below.

The Normal distribution

If a random variable $X \in \mathbb{R}$ has density function

$$f(x) = \frac{1}{\sqrt{2\pi}\sigma} e^{-\frac{(x-\mu)^2}{2\sigma^2}}, \tag{2.17}$$

then it is said to follow the $\text{N}(\mu, \sigma^2)$ distribution, with $\mathbb{E}(X) = \mu$ and $\mathbb{V}\text{ar}(X) = \sigma^2$ (see Exercise 2.10), and the standardized random variable

$$Z \equiv \frac{X - \mu}{\sigma} \sim \text{N}(0, 1)$$

is said to follow the *standard normal distribution* (see Exercise 3.6 in section 3.2.1). Equation (2.17) is famously known as the "bell curve" due to its shape; see Figure 2.4.

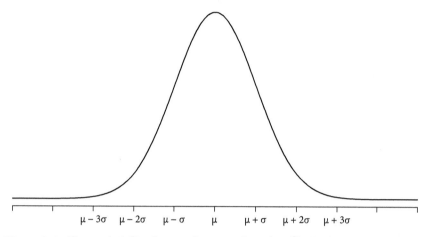

Figure 2.4   The probability density function for $N(\mu, \sigma^2)$, the "bell curve".
Source: authors.

Undoubtedly a big reason why the normal distribution is so important is because of the *central limit theorem*, which says that, if you add (or average) enough random quantities independently drawn from the same distribution (almost any distribution), the quantity you obtain will have approximately a bell curve distribution!

Theorem 1 (Central limit theorem). *Suppose $X_1, \ldots, X_n$ are independent and identically distributed (i.i.d.) random variables with $\mathbb{E}(X_i) = \mu$ and $\mathbb{V}\mathrm{ar}(X_i) = \sigma^2 < \infty$. Let $S_n = X_1 + X_2 + \cdots + X_n$ be their sum. Then, as $n \to \infty$, the distribution of*

$$\frac{S_n - n\mu}{\sqrt{n}\sigma}$$

converges to that of $N(0, 1)$. $\qquad\qquad\qquad\qquad\qquad\qquad\square$

Remark 2.5.   Let $\bar{X}_n = S_n/n$ be the empirical average of these n random variables. Then, it is clear in Theorem 1 above that one can divide both the numerator and the denominator by n to obtain the same result for

$$\frac{\bar{X}_n - \mu}{\sigma/\sqrt{n}}.$$

Thus, the central limit theorem applies equally to *averages* of i.i.d. random variables, even though the statement appears to be only about *sums*; the only difference is that the sum (S_n) and the average ($\bar{X}_n$) must be shifted and scaled by different amounts in order for the standard normal conclusion to hold for them both.

Based on the central limit theorem, we often say that S_n is approximately distributed as $N(n\mu, n\sigma^2)$, and $\bar{X}_n$ is approximately distributed as $N(\mu, \sigma^2/n)$. □

Exercise 2.9. Show that

$$\int e^{-\frac{z^2}{2}} dz = \sqrt{2\pi}.$$

[*Hint: Let $A = \int e^{-\frac{z^2}{2}} dz$. Compute*

$$A^2 = \left(\int e^{-\frac{x^2}{2}} dx\right)\left(\int e^{-\frac{y^2}{2}} dy\right) = \iint e^{-\frac{x^2+y^2}{2}} dxdy$$

by using polar coordinates.] □

Exercise 2.10. Suppose $X \sim N(\mu, \sigma^2)$. Show that $\mathbb{E}(X) = \mu$ and $\mathbb{V}\text{ar}(X) = \sigma^2$. [*Hint: You may want to refresh yourself on the two basic integration techniques: change of variables, and integration by parts.*] □

Exercise 2.11. Suppose we have a measurement:

$$X \sim \begin{cases} N(\mu_0, \sigma_0^2), & \text{if the person being measured is healthy } (H) \\ N(\mu_1, \sigma_1^2), & \text{if the person is sick } (S), \end{cases}$$

where $\mu_1 > \mu_0$. Based on such a measurement, the person will be diagnosed as being sick if $X \geq c$, for some decision threshold c.

(a) Assume that $\mathbb{P}(H) = \mathbb{P}(S) = 1/2$ and $\sigma_0^2 = \sigma_1^2$. Find the optimal threshold that minimizes the probability of an erroneous diagnosis,

$$\mathbb{P}(\text{error}) = \mathbb{P}(X \geq c|H)\mathbb{P}(H) + \mathbb{P}(X < c|S)\mathbb{P}(S).$$

(b) How will the optimal threshold change if $\mathbb{P}(H) \neq \mathbb{P}(S)$ and/or if $\sigma_0^2 \neq \sigma_1^2$?

[*Hint: Let $F_0(x)$, $F_1(x)$ be the cumulative distribution functions of the two normal distributions, respectively. Express $\mathbb{P}(\text{error})$ in terms of $F_0(c)$ and $F_1(c)$; then, differentiate with respect to c.*] □

3

Gaining Fluency

In real problems, we must almost always deal with multiple random quantities. This chapter helps students to become more fluent at the language of probability by developing the important skill for doing so. When we have more than one random variable, even the very idea of the distribution itself becomes richer as there are different kinds of distributions for us to talk about.

3.1 Multiple random quantities

First, we will use a simple example to illustrate some main principles.

Example 3.1. Let's go back to the very simple case of tossing two dice, considered earlier in Examples 2.1 and 2.2. In this case, we can define two random variables: $X \in \{1, 2, \ldots, 6\}$ for the outcome of the first die and $Y \in \{1, 2, \ldots, 6\}$ for that of the second. The *joint distribution* of (X, Y), here described by their joint probability mass function,

$$f(x, y) = \mathbb{P}[(X = x) \cap (Y = y)],$$

is simply a 6×6 table, listing the probabilities of all $6 \times 6 = 36$ outcomes, as shown below.

$x \backslash y$	1	2	$\ldots$	6
1	p_{11}	p_{12}	$\cdots$	p_{16}
2	p_{21}	p_{22}	$\cdots$	p_{26}
$\vdots$	$\vdots$	$\vdots$	$\ddots$	$\vdots$
6	p_{61}	p_{62}	$\cdots$	p_{66}

Essential Statistics for Data Science. Mu Zhu, Oxford University Press. © Mu Zhu (2023).
DOI: 10.1093/oso/9780192867735.003.0003

How would we compute a quantity like $\mathbb{P}(X + Y = 4)$? We simply pick out all pairs of (x, y) such that $x + y = 4$—namely, $(1, 3), (2, 2), (3, 1)$—and add up their respective probabilities, that is,

$$\mathbb{P}(X + Y = 4) = f(1, 3) + f(2, 2) + f(3, 1) = \sum_{x+y=4} f(x, y) = p_{13} + p_{22} + p_{31}.$$

If the dice are fair (unloaded), and tossed independently, then all 36 outcomes are equally likely, each with probability $1/36$, and $\mathbb{P}(X + Y = 4) = 3/36$. Here, we are being a bit more general, allowing the possibility that the dice may be loaded, so the 36 entries may not all have the same probability. □

This simple example nonetheless shows us a general principle: If we want to compute the probability of an event characterized by both X and Y, such as $g(X, Y) \in A \subset \mathbb{R}$, we simply pick out all pairs of (x, y), such that $g(x, y) \in A$, and aggregate their probabilities. Again, we use the word "aggregate" deliberately to mean both "sum" and "integrate", depending on whether the underlying random variables are discrete or continuous, so

$$\mathbb{P}[g(X, Y) \in A] \quad = \quad \sum_{g(x,y) \in A} f(x, y) \quad \text{or} \quad \int_{g(x,y) \in A} f(x, y) \, dxdy. \quad (3.1)$$

Example 3.2. Let us continue with Example 3.1. If we only care about the outcome of the second die (e.g. the event $\{Y = 4\}$), then we simply pick out all pairs of (x, y), such that $y = 4$, and add up their probabilities, that is,

$$\mathbb{P}(Y = 4) = f(1, 4) + f(2, 4) + \ldots + f(6, 4) = \sum_{x} f(x, 4).$$

That is, we simply fix $f(x, y)$ at $y = 4$ and sum over all possible values of x. This makes sense because we only care about the event $\{Y = 4\}$ and do not care what X is, so X can be anything. □

This simple operation shows us another general principle: To compute the *marginal distribution* of one variable, simply aggregate over the other, so

$$f_X(x) \quad = \quad \sum_{y} f(x, y) \quad \text{or} \quad \int f(x, y) \, dy, \quad (3.2)$$

and likewise for $f_Y(y)$.

Finally, we can also talk about the *conditional distribution* of one, given the other. This is defined analogously to the conditional probability (Definition 1 in section 2.2), that is,

$$f_{Y|X}(y|x) \;=\; \frac{f(x,y)}{f_X(x)}, \tag{3.3}$$

and likewise for $f_{X|Y}(x|y)$. The conditional distribution is the one we would focus on if we would like to predict, say, Y from X.

As before, if Y is *independent* of X, then its conditional and unconditional distributions must be the same; that is, $f_{Y|X}(y|x) = f_Y(y)$, meaning their joint distribution must factor, $f(x,y) = f_X(x)f_Y(y)$.

Exercise 3.1.   Amy and Bob agree to meet at the library to work on homework together. Independently, they are both equally likely to arrive at any time between 7 pm and 8 pm. What is the probability that Amy has to wait at least 15 minutes for Bob? [*Hint: Let X and Y be independent random variables, each distributed uniformly on the interval* $(0, 1)$. *Find* $\mathbb{P}(X - Y > 1/4)$.] $\square$

Example 3.3.   Let us first look at a simple, discrete, bivariate distribution with joint probability mass function,

$$f(x,y) = \frac{n!}{x!y!(n-x-y)!}p_1^x p_2^y (1-p_1-p_2)^{n-x-y}, \quad x,y \in \{0,1,\ldots,n\}.$$

This is a special case of the Multinomial$(n; p_1, p_2, \ldots, p_K)$ distribution—a generalization of the Binomial(n, p) distribution with joint mass function,

$$f(x_1, x_2, \ldots, x_K) = \frac{n!}{x_1! x_2! \ldots x_K!} p_1^{x_1} p_2^{x_2} \cdots p_K^{x_K},$$

$$\text{s.t.} \quad \sum_{k=1}^{K} p_k = 1 \quad \text{and} \quad \sum_{k=1}^{K} x_k = n.$$

If the binomial distribution describes n independent coin flips (two possible outcomes on each flip), the multinomial distribution describes n independent tosses of a K-faced die (K possible outcomes on each toss). Due to the constraint, $X_1 + \ldots + X_K = n$, a K-category multinomial is really a $(K-1)$-dimensional distribution. The case considered here corresponds to $K = 3$.

To obtain the marginal distribution of X, we sum over all possible values of y:

$$f_X(x) = \sum_{y=0}^{n-x} \frac{n!}{x!y!(n-x-y)!} p_1^x p_2^y (1-p_1-p_2)^{n-x-y}$$

$$= \frac{n!}{x!(n-x)!} p_1^x \underbrace{\sum_{y=0}^{n-x} \frac{(n-x)!}{y!(n-x-y)!} p_2^y (1-p_1-p_2)^{n-x-y}}_{(p_2+1-p_1-p_2)^{n-x}}$$

$$\overset{\star}{=} \frac{n!}{x!(n-x)!} p_1^x (1-p_1)^{n-x}$$

$$\sim \text{Binomial}(n, p_1),$$

where the step marked by "$\star$" uses the Binomial theorem (see Mathematical Insert 1 below).

A subtle detail here is the summation limits. Students often get this wrong on their first try. At first, it seems summing over all possible values of y means that we should sum from 0 all the way up to n. We will leave it as an exercise (Exercise 3.2) for you to think why we should sum only up to $n - x$ instead.

The conditional distribution of Y given $X = x$, then, is simply

$$f_{Y|X}(y|x) = \frac{\left[\frac{n!}{x!y!(n-x-y)!} p_1^x p_2^y (1-p_1-p_2)^{n-x-y} \right]}{\left[\frac{n!}{x!(n-x)!} p_1^x (1-p_1)^{n-x} \right]}. \qquad (3.4)$$

We will leave it as an exercise (Exercise 3.2) for you to simplify Equation (3.4) and discover what kind of distribution it actually is. $\qquad \square$

Mathematical Insert 1

Binomial theorem.

$$(a+b)^m = \sum_{i=0}^{m} \frac{m!}{i!(m-i)!} a^i b^{m-i}.$$

Exercise 3.2. Complete Example 3.3.

(a) To obtain the marginal distribution of X, why do we sum from $y = 0$ up to $y = n - x$ only?

(b) Simplify Equation (3.4) and show that the conditional distribution of Y, given $X = x$, is Binomial $\left(n - x, \frac{p_2}{1 - p_1}\right)$.

[*Think: Why does the conclusion in (b) make sense intuitively?*]   □

Example 3.4. Let us now look at a simple, continuous, bivariate distribution with joint probability density function,

$$f(x, y) = \frac{1}{2\pi\sqrt{1 - \rho^2}} \exp\left[-\frac{x^2 - 2\rho xy + y^2}{2(1 - \rho^2)}\right], \quad x, y \in \mathbb{R}, \quad \rho \in (-1, 1).$$

This is a special case of the n-dimensional Normal(μ, Σ) distribution,

$$f(z) = \frac{1}{\sqrt{(2\pi)^n |\Sigma|}} \exp\left[-\frac{(z - \mu)^\mathsf{T} \Sigma^{-1}(z - \mu)}{2}\right], \quad z \in \mathbb{R}^n, \tag{3.5}$$

with $n = 2$,

$$z = \begin{bmatrix} x \\ y \end{bmatrix} \in \mathbb{R}^2, \quad \mu = \begin{bmatrix} 0 \\ 0 \end{bmatrix}, \quad \Sigma = \begin{bmatrix} 1 & \rho \\ \rho & 1 \end{bmatrix}.$$

To obtain the marginal distribution of X, we integrate over all y, and obtain

$$
\begin{aligned}
f_X(x) &= \int f(x, y)\, dy \\
&\overset{\star}{=} \int \frac{1}{2\pi\sqrt{1 - \rho^2}} \exp\left[-\frac{y^2 - 2\rho xy + \rho^2 x^2 + (1 - \rho^2)x^2}{2(1 - \rho^2)}\right] dy \\
&= \frac{1}{\sqrt{2\pi}} e^{-\frac{x^2}{2}} \times \underbrace{\int \frac{1}{\sqrt{2\pi(1 - \rho^2)}} \exp\left[-\frac{(y - \rho x)^2}{2(1 - \rho^2)}\right] dy}_{N(\rho x, 1 - \rho^2)} \\
&\sim N(0, 1).
\end{aligned}
$$

The step marked by "$\star$" is a standard "trick" for dealing with normal distributions. Because the density function of a normal distribution is log-quadratic, the "trick" is to complete the square in the exponent so that the integrand can be recognized as a certain normal density function, in which case it will automatically integrate to one.

Note that this "trick" immediately implies that the conditional distribution of Y, given $X = x$, is $N(\rho x, 1 - \rho^2)$. Why? Because in essence we first factored $f(x, y)$ into

$$f(x, y) = h(x)g(x, y)$$

and then pulled $h(x)$ out of the integral. What happened next? We noticed that $\int g(x, y)dy = 1$, so $h(x) = f_X(x)$ is, in fact, the marginal distribution of X. But this means the component remaining inside the integral,

$$g(x, y) = \frac{f(x, y)}{h(x)} = \frac{f(x, y)}{f_X(x)},$$

must be $f_{Y|X}(y|x)$ by definition. $\qquad\qquad\qquad\qquad\qquad\qquad\square$

Remark 3.1. The multinomial distribution and the multivariate normal distribution are perhaps the two most ubiquitous probability models that we encounter in practice—one for categorical data and one for real-valued data. Examples 3.3 and 3.4 gave us a first (and only) taste of these models in this book by offering a glimpse of their simplest cases—namely, a trinomial and a bivariate normal.

Nonetheless, we can already see that the sums and integrals in Equations (3.1) and (3.2) are not easy to handle. We often need special knowledge (e.g. the Binomial theorem for Example 3.3) and special tricks (e.g. "completing the square inside the exponent" for Example 3.4), and there can be subtle details (e.g. the summation limits in Example 3.3) to throw us off the track, too.

While it is important to appreciate both the difficulty and the importance of these types of calculations (also see section 3.1.1 below), in this book we will try to stay away as much as possible from having to learn specific tricks to deal with different distributions, so that we can focus on the main ideas and principles instead. $\qquad\qquad\qquad\qquad\qquad\qquad\square$

Remark 3.2. If we want to predict Y based on X, it is natural—and, in some sense, optimal—to use the *conditional expectation* of Y given X, that is, the expectation of the conditional distribution $f_{Y|X}(y|x)$. This is another very important idea in probability but one which we will again mostly skip in this book. We will simply state what this means for the two examples we have considered above. For the trinomial (Example 3.3), Exercise 3.2(b) above implies that, based on X, we may predict Y to be $(n - X)(p_2/(1 - p_1))$ (see also Exercise 3.11 in section 3.A.3). For the bivariate normal (Example 3.4), it means that, based on X, we may predict Y to be ρX. $\quad\square$

Exercise 3.3. Two positive random variables, X and Y, have joint probability density function given by

$$f(x, y) = ye^{-y(x+1)}, \qquad x, y > 0.$$

(a) Find the marginal distribution of Y, $f_Y(y)$.
(b) Find the conditional distribution of X given Y, $f_{X|Y}(x|y)$.
(c) If you are told that $Y = 2$, what do you expect X to be, on average?

[*Note: What about the other way round? What is the conditional distribution of Y, given X, $f_{Y|X}(y|x)$? The underlying principle is the same, of course, but the actual calculation may become easier after one has learned about the so-called Gamma distribution (see section 4.3.2).*] $\quad\square$

3.1.1 Higher-dimensional problems

If we have more than just two random variables, say, $X_1, X_2, \ldots, X_n$, the ideas are exactly the same, but the calculations will become more tedious and a lot harder. In particular, Equation (3.1) will become high-dimensional sums and/or integrals,

$$\sum_{g(x_1, \ldots, x_n) \in A} f(x_1, \ldots, x_n) \text{ or } \int_{g(x_1, \ldots, x_n) \in A} f(x_1, \ldots, x_n) \, dx_1 \ldots dx_n. \quad (3.6)$$

That is, we pick out all n-tuples $(x_1, \ldots, x_n)$, such that $g(x_1, \ldots, x_n) \in A$, and aggregate their probabilities.

More often than not, these (high-dimensional) sums and integrals will be intractable. It is not an exaggeration to say that if you have a *general* way to compute these high-dimensional sums and/or integrals, then you can claim to have solved many research problems in data science and machine learning. One can certainly characterize a lot of current research as being about how to specify the probability model so that the joint distribution $f(x_1, \ldots, x_n)$ can both represent reality and have certain special structures allowing Equation (3.6) to be computed (or reasonably well approximated) with some specialized "tricks".

This does *not* mean, however, that problems involving just two random variables are all "toy" problems. In fact, when dealing with a high-dimensional random vector, we often partition it into two blocks. For example, for $Z \in \mathbb{R}^{1000}$, we may partition it into a vector in $\mathbb{R}^{800}$ and another one in $\mathbb{R}^{200}$,

$$
Z = \begin{bmatrix} Z_1 \\ \vdots \\ Z_{800} \\ \hline Z_{801} \\ \vdots \\ Z_{1000} \end{bmatrix} = \begin{bmatrix} X \\ \hline Y \end{bmatrix}, \quad X \in \mathbb{R}^{800}, Y \in \mathbb{R}^{200},
$$

and still think as if there were just two (multivariate) random quantities, X and Y, with joint distribution $f(x, y)$; marginal distributions $f_X(x)$, $f_Y(y)$; and conditional distributions $f_{Y|X}(y|x)$, $f_{X|Y}(x|y)$. Hence, mastering the case of two random variables is not just a baby step but a big one toward being able to handle higher-dimensional problems.

Example 3.5. If $Z \in \mathbb{R}^d$ follows the multivariate $N(\mu, \Sigma)$ distribution with density function given by Equation (3.5), then a partition of Z into two blocks, together with a corresponding partition of the two (vector and matrix) parameters,

$$
Z = \begin{bmatrix} X \\ Y \end{bmatrix}, \quad \mu = \begin{bmatrix} \mu_1 \\ \mu_2 \end{bmatrix}, \quad \Sigma = \begin{bmatrix} \Sigma_{11} & \Sigma_{12} \\ \Sigma_{21} & \Sigma_{22} \end{bmatrix},
$$

will allow us to derive (details omitted, of course) that

$$
Y|X \sim N(\mu_2 + \Sigma_{21}\Sigma_{11}^{-1}(X - \mu_1), \Sigma_{22} - \Sigma_{21}\Sigma_{11}^{-1}\Sigma_{12}).
$$

In fact, Example 3.4 is merely a very special case of this more general problem. (Verify it.) Thus, based on X, we may use $\mu_2 + \Sigma_{21}\Sigma_{11}^{-1}(X - \mu_1)$, a linear function of X, to predict Y. (See also Remark 3.2 earlier in this section.) $\square$

3.2 Two "hard" problems

Dealing with multiple random quantities is hard. At the intuitive level, the reason why randomness is hard is because things are changing, and we are not sure what will happen. If things don't change, and we always know what will happen, then life is easy. So, when *many* things are changing at the same time, it's easy to understand how we can quickly lose our sanity.

In order to proceed with our study of statistics, it is necessary to become somewhat comfortable with two specific problems in probability, both of which are somewhat challenging for students precisely because they involve multiple random quantities, and it takes time and experience before even the best of students can master dealing with multiple sources of randomness. In this chapter, we simply focus on these two problems on their own. We will explain why they are relevant to statistics later, in Chapter 4.

3.2.1 Functions of random variables

The first problem can be described as follows. Suppose we know the joint distribution of n random variables, say, $X_1, X_2, \ldots, X_n$, and we are interested in a particular function of them, say, $Y = g(X_1, X_2, \ldots, X_n)$. For example, $X_1, X_2, \ldots, X_n$ may be the respective amount of time it takes for a claim to come forward from n independent insurance policies sold by a certain broker, and Y may be how long this broker can stay "claim free". [*Think: How could we express Y as a function of $X_1, X_2, \ldots, X_n$ in this case?*] Clearly, the quantity Y is also random—unless the function g is trivial, such as $g(\cdot) \equiv c$, where c is a (non-random) constant. Since Y is random, it is natural for us to ask how it is distributed.

To answer this question, the basic idea here is that we now know how to calculate the probability of an event like $g(X_1, X_2, \ldots, X_n) \le y$, for any given y. Do you remember what we said earlier about the situation being analogous to tossing two dice? We just have to look for all the combinations that meet the criterion specified by the event and aggregate—either sum or integrate— over all those eligible combinations; see, for example, Exercise 3.1. (We will focus on the case in which all random variables involved are continuous.) This means we know, in principle, how to calculate the cumulative distribution function of Y,

$$F_Y(y) = \mathbb{P}(Y \le y) = \int_{g(x_1, \ldots, x_n) \le y} f(x_1, \ldots, x_n) \, dx_1 \ldots dx_n. \qquad (3.7)$$

We can then differentiate and obtain the density function of Y:

$$f_Y(y) = \frac{d}{dy} F_Y(y).$$

We say "in principle" because this task is more easily said than done. More specifically, the integral in Equation (3.7) above is n-dimensional, and evaluating it will be very hard in general. Let's first look at a relatively easy example in order to see how this is actually done.

Example 3.6. Suppose that we have two independent random variables, X and Y, both following the Uniform$(0, 1)$ distribution, and that we are interested in their sum, $Z = X + Y$. From the basic principle, we get

$$
\begin{aligned}
F_Z(z) &= \mathbb{P}(Z \leq z) \\
&= \mathbb{P}(X + Y \leq z) \\
&= \int_{x+y \leq z} f_X(x) f_Y(y)\, dxdy \\
&= \int_{x+y \leq z} 1\, dxdy.
\end{aligned}
$$

Now, we see that what makes this example particularly easy is that, even though we must evaluate a double integral, the function we need to integrate is merely a constant.

Whereas a one-dimensional integral computes the area under the curve, so a two-dimensional integral computes the volume under the surface. Here, we have a flat surface, $f(x, y) = 1$, so the volume can be computed with the elementary formula,

$$(\text{volume}) = (\text{base area}) \times (\text{height}),$$

with (height) $= 1$. Thus, by Figure 3.1, we get

$$
F_Z(z) = \begin{cases}
\dfrac{z^2}{2}, & z \leq 1; \\
1 - \dfrac{(2-z)^2}{2}, & z > 1.
\end{cases}
$$

Of course, the sum of two quantities between 0 and 1 must be between 0 and 2, so outside this range the function is trivial: $F_Z(z) = 0$ for $z < 0$ and $F_Z(z) = 1$ for $z > 2$. This gives

$$
f_Z(z) = \frac{d}{dz} F_Z(z) = \begin{cases}
z, & 0 < z \leq 1; \\
2 - z, & 1 < z < 2.
\end{cases}
$$

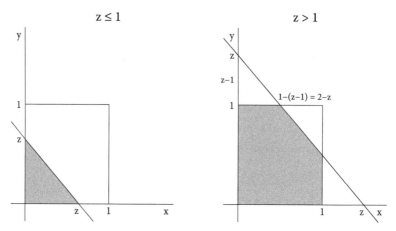

Figure 3.1 The region, $\{(x, y) : x + y \leq z, (x, y) \in (0, 1) \times (0, 1)\}$, looks slightly different depending on whether $z \leq 1$ or $z > 1$.
Source: authors.

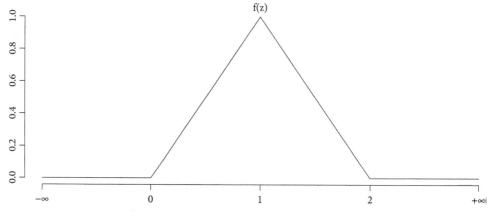

Figure 3.2 The density function of $Z = X + Y$ in Example 3.6.
Source: authors.

The resulting density (Figure 3.2) is triangular-shaped on $(0, 2)$, with a peak at $z = 1$. What this says is that it is more likely for the random variable Z to be near 1 than for it to be near 0 or 2. [*Think: Does this make sense?*] ☐

But, as we stated prior to Example 3.6, evaluating the integral in Equation (3.7) can be very hard in general. In fact, that was a deliberate understatement! Evaluating such an integral in closed form is often impossible—unless in special cases. Let's look at one of these special cases next, the case of $g(X_1, X_2, \ldots, X_n) = \max\{X_1, X_2, \ldots, X_n\}$.

Example 3.7. Suppose that we have a set of independent random variables, say $X_1, \ldots, X_n$, each with density function $f_1(x), \ldots, f_n(x)$, respectively, and that we are interested in their maximum, $Y = \max\{X_1, \ldots, X_n\}$.

What makes this case very special is that the probability, $\mathbb{P}(Y \leq y)$, can be computed by completely avoiding the intimidating n-dimensional integral in Equation (3.7). Needless to say, this does not happen as often as we would like.

Notice that $Y \leq y$ if, and only if, all of $X_1, X_2, \ldots, X_n$ are no larger than y.[1] Thus, the event $\{Y \leq y\}$ has the same probability as the event $\{$all of $X_1, X_2, \ldots, X_n \leq y\}$, which means

$$
\begin{aligned}
F_Y(y) &= \mathbb{P}(Y \leq y) \\
&= \mathbb{P}[(X_1 \leq y) \cap (X_2 \leq y) \cap \ldots \cap (X_n \leq y)] \\
&= \prod_{i=1}^{n} \mathbb{P}(X_i \leq y) \quad \text{(due to independence)} \\
&= \prod_{i=1}^{n} F_i(y),
\end{aligned}
$$

where each $F_i(y) = \int_{-\infty}^{y} f_i(x)dx$ is simply the corresponding cumulative distribution function of X_i, for $i = 1, 2, \ldots, n$.

This special case simplifies further if the n random variables above are not only independent but also identically distributed—that is, if $f_i(x) = f_X(x)$ and $F_i(x) = F_X(x)$ for all $i = 1, 2, \ldots, n$. Then,

$$
F_Y(y) = [F_X(y)]^n \quad \Rightarrow \quad f_Y(y) = \frac{d}{dy} F_Y(y) = n[F_X(y)]^{n-1} f_X(y).
$$

As an even more specific example, consider the simple case in which $f_X(x)$ is the Uniform$(0, 1)$ density function. This means that, on the interval $(0, 1)$,

$$
f_X(x) = 1 \quad \text{and} \quad F_X(x) = \int_0^x f_X(t)dt = \int_0^x 1 dt = x.
$$

Thus, their maximum Y has density function

$$
f_Y(y) = ny^{n-1}, \quad y \in (0, 1).
$$

---

[1] Convince yourself of the "if and only if" here. If $X_i \leq y$ for all $i = 1, 2, \ldots, n$, then $Y = \max\{X_1, X_2, \ldots, X_n\} \leq y$. On the other hand, if at least one of $X_1, \ldots, X_n$ is larger than y, then $Y = \max\{X_1, X_2, \ldots, X_n\}$ must be larger than y as well.

It is easy to see that this is an increasing function of y, which means values closer to 1 are more likely for Y than those closer to 0. [*Think: Does this make sense?*] With the density function at hand, we can also easily compute the expectation of Y:

$$\mathbb{E}(Y) = \int_0^1 y f_Y(y) dy = \int_0^1 y(ny^{n-1}) dy = n \left[\frac{y^{n+1}}{n+1} \right]_0^1 = \frac{n}{n+1} \rightarrow 1 \text{ as } n \rightarrow \infty.$$

[*Think: Does it makes sense that the maximum of n independent random variables from the Uniform(0, 1) distribution should be equal to n/(n + 1) on average?*] □

Another special case that we encounter very often is the case of the sum, for example,

$$g(X_1, X_2, \ldots, X_n) = \sum_{i=1}^n a_i X_i, \tag{3.8}$$

where $a_1, a_2, \ldots, a_n$ are (non-random) constants, and $X_1, X_2, \ldots, X_n$ are independent. Because of its ubiquity, we will devote a separate appendix (Appendix 3.A at the end of this chapter) to it. While various specific results in Appendix 3.A will be important for practical reasons, they do not necessarily add to our conceptual understanding of what the distribution of $g(X_1, X_2, \ldots, X_n)$ means or how to work it out in principle.

Exercise 3.4. Suppose that we have a set of independent random variables, say $X_1, \ldots, X_n$, each with density function $f_1(x), \ldots, f_n(x)$, respectively, and that we are interested in their minimum, $Z = \min \{X_1, \ldots, X_n\}$.

(a) Assume that $f_i(x) = f_X(x)$ for all $i = 1, 2, \ldots, n$. Find the density function of Z.
(b) Further, assume that $f_X(x)$ is the Uniform(0, 1) density function. Find $\mathbb{E}(Z)$.

[*Hint: For part (b), you will need to know about the so-called Beta function; see Mathematical Insert 2.*] □

Exercise 3.5. Given a standard normal random variable $Z \sim N(0, 1)$, whose density function is

$$f_Z(z) = \frac{1}{\sqrt{2\pi}} e^{-z^2/2},$$

Mathematical Insert 2

Beta function. In mathematics, the Beta function is defined as

$$Be(\alpha, \beta) = \int_0^1 t^{\alpha-1}(1-t)^{\beta-1}dt = \frac{\Gamma(\alpha)\Gamma(\beta)}{\Gamma(\alpha+\beta)},$$

where $\Gamma(\cdot)$ is the so-called Gamma function—its definition and some of its very useful properties are given in Mathematical Insert 3.

Mathematical Insert 3

Gamma function. The Gamma function is defined as

$$\Gamma(\alpha) = \int_0^\infty t^{\alpha-1}e^{-t}dt,$$

and it has the following properties:

(a) $\Gamma(\alpha + 1) = \alpha\Gamma(\alpha)$;
(b) $\Gamma(1) = 1$;
(c) $\Gamma(1/2) = \sqrt{\pi}$.

Note that (a) and (b) make it clear that $\Gamma(\alpha) = (\alpha - 1)!$ if α is an integer, so the Gamma function can be regarded as a generalized factorial.

---

find the density function of $X = Z^2$. [*Note: The resulting distribution of X is known as the "chi-squared distribution with one degree of freedom", often denoted as $\chi^2_{(1)}$.*] □

Exercise 3.6. Given a standard normal random variable $Z \sim N(0, 1)$, find the density function of $X = \sigma Z + \mu$, where $\sigma > 0$. □

3.2.2 Compound distributions

The second problem is concerned with the following. Suppose we have a random variable X, which follows a certain distribution $f(x; \theta)$, with parameter θ. What if the parameter θ itself is also random, with distribution, say, $\pi(\theta)$? For example, X may be the number of sales at a certain car dealership today, and θ may be the total number of customers. (See Exercise 4.2 in section 4.3.1.)

In this case, the distribution $f(x; \theta)$ is really the *conditional* distribution of X, given θ, $f(x|\theta)$, because, when describing the behavior of X, we are pretending that θ is not random whereas, in fact, it is. Together with $\pi(\theta)$, we have the *joint* distribution of (X, θ),

$$f(x, \theta) = f(x|\theta)\pi(\theta),$$

from which we can then obtain the *marginal* distribution of X by aggregating over all possible values of θ. That is,

$$f(x) = \sum_{\theta} f(x, \theta) = \sum_{\theta} f(x|\theta)\pi(\theta) \tag{3.9}$$

or

$$f(x) = \int f(x, \theta) d\theta = \int f(x|\theta)\pi(\theta) d\theta, \tag{3.10}$$

depending on whether θ is discrete or continuous.

Under such circumstances, the resulting marginal distribution of X, after having aggregated over all possible values of the parameter θ, is sometimes referred to as a *compound distribution*. Let's look at a relatively simple example to see what all this means.

Example 3.8.   Suppose that $X \sim \text{Binomial}(n, p)$, where the parameter p is also random, say, $p \sim \text{Uniform}(0, 1)$. Then, for each $x \in \{0, 1, 2, \ldots, n\}$,

$$f(x) = \mathbb{P}(X = x)$$

$$= \int_0^1 \underbrace{\mathbb{P}(X = x|p)}_{\text{Binomial}(n, p)} \times \underbrace{\pi(p)}_{\text{Uniform}(0,1)} dp$$

$$= \int_0^1 \left[\binom{n}{x} p^x (1 - p)^{n-x} \right] \times (1) \, dp$$

$$= \binom{n}{x} \int_0^1 p^x (1 - p)^{n-x} \, dp$$

$$\stackrel{\star}{=} \frac{n!}{x!(n-x)!} \times \frac{\Gamma(x+1)\Gamma(n-x+1)}{\Gamma(n+2)}$$

$$\stackrel{\star\star}{=} \frac{n!}{x!(n-x)!} \times \frac{x!(n-x)!}{(n+1)!}$$

$$= \frac{1}{n+1},$$

where the step marked by "$\star$" is due to how the Beta function is defined (see Mathematical Insert 2 in section 3.2.1), and the step marked by "$\star\star$" uses properties of the Gamma function (see Mathematical Insert 3 in section 3.2.1).

Thus, each of the $n+1$ possible values $\{0, 1, 2, \ldots, n\}$ is equally likely to occur with probability $1/(n+1)$. This is clearly very different from the Binomial distribution, which peaks near np (see Figure 2.3 in section 2.3.3). Intuitively, this is easily understood because now the parameter p is no longer fixed; instead, it is equally likely to be any value between 0 and 1, so the Binomial peak near np is "washed out" and any integer value between 0 and n becomes equally likely.

What happens if the parameter n is a random variable instead? Since n is discrete rather than continuous, Equation (3.9) is applicable instead of Equation (3.10), so

$$f(x) = \mathbb{P}(X = x) = \sum_n \mathbb{P}(X = x|n)\pi(n),$$

where $\pi(n)$ is the probability mass function of n. Exercise 4.2 (in section 4.3.1) contains a special case in which the sum above can be computed in closed form. □

From Example 3.8, we can see that the main difficulty here is similar to the one we encountered earlier in section 3.2.1; generally, the integral in Equation (3.10) is by no means easy to evaluate—in fact, it is often impossible to evaluate in closed form, unless in very special cases. The same can be said of the sum in Equation (3.9).

Exercise 3.7. Suppose that, in Example 3.8, there is another random variable $X_{new} \sim$ Binomial$(1, p)$ that is *conditionally independent* of X, given p, that is,

$$\mathbb{P}(X_{new} = 1 \text{ and } X = x|p) = \mathbb{P}(X_{new} = 1|p) \times \mathbb{P}(X = x|p).$$

Find

$$\mathbb{P}(X_{new} = 1 \text{ and } X = x) =$$

$$\int \underbrace{\mathbb{P}(X_{new} = 1|p)}_{\text{Binomial}(1,p)} \times \underbrace{\mathbb{P}(X = x|p)}_{\text{Binomial}(n,p)} \times \underbrace{\pi(p)}_{\text{Uniform}(0,1)} dp$$

and show that $\mathbb{P}(X_{new} = 1|X = x) = (x+1)/(n+2)$. [*Think: This result is known as "Laplace's rule of succession". Why does a larger value of X make the event $X_{new} = 1$ more likely even though the two random variables are conditionally independent?*] □

Appendix 3.A Sums of independent random variables

It is not hard to understand why Equation (3.8), with $X_1, X_2, \ldots, X_n$ being independent of one another, contains perhaps the most frequently encountered function of n random variables. As such, there are many special techniques and results for dealing with functions of this type.

3.A.1 Convolutions

We have already seen (e.g. Example 3.6 in section 3.2.1) that, when $n = 2$, we must evaluate a double integral in order to get the answer, but sometimes this double integral can be "pre-reduced".

For example, suppose $Z = X + Y$, where X and Y are two independent random variables. Starting from first principles, we can derive

$$
\begin{aligned}
F_Z(z) &= \mathbb{P}(Z \leq z) \\
&= \mathbb{P}(X + Y \leq z) \\
&= \int_{x+y \leq z} f_X(x) f_Y(y) dx dy \\
&= \int_{\text{all } x} \int_{y \leq z-x} f_X(x) f_Y(y) dy dx \\
&= \int_{\text{all } x} f_X(x) \left[\int_{y \leq z-x} f_Y(y) dy \right] dx \\
&= \int_{\text{all } x} f_X(x) F_Y(z - x) dx
\end{aligned}
$$

and

$$f_Z(z) = \frac{d}{dz} F_Z(z) \overset{\star}{=} \int_{\text{all } x} f_X(x) \left[\frac{d}{dz} F_Y(z-x) \right] dx$$

$$= \int_{\text{all } x} f_X(x) f_Y(z-x) dx, \tag{3.11}$$

where we have interchanged the order of differentiation and integration[2] in the step marked by "$\star$". Equation (3.11) is called a *convolution*.

Remark 3.3. In practice, one needs to be careful with the integration region generically denoted by "all x" in (3.11). For example, if $X, Y \in \mathbb{R}$, then "all x" will mean $x \in \mathbb{R}$; but if $X, Y > 0$, then "all x" will mean $0 < x < z$. ☐

Exercise 3.8. Suppose $X \sim N(\mu_1, \sigma_1^2)$ and $Y \sim N(\mu_2, \sigma_2^2)$ are two independent random variables. Use the convolution formula (3.11) to find the density function of $Z = X + Y$. [*Note: If you wish, assuming $\sigma_1^2 = \sigma_2^2 = \sigma^2$ will surely make the algebra much more bearable.*] ☐

3.A.2 Moment-generating functions

Another special device that is useful for dealing with (3.8), if all $X_1, X_2, \ldots, X_n$ are independent, is the *moment-generating function*. Given a random variable X, its moment-generating function is defined as

$$m_X(t) = \mathbb{E}(e^{tX}) \quad \left[\; = \int e^{tx} f_X(x) dx \quad \text{or} \quad \sum_x e^{tx} f_X(x) \; \right]. \tag{3.12}$$

In spirit, this device is similar to the *Laplace transform* or the *Fourier transform*—the main reason why they are useful is because sometimes problems that are hard to solve directly can be made easier if we first apply such a transform, solve the problem in the transformed domain, and then apply the inverse transform back to the original domain. Here, the problem of finding the distribution of $a_1 X_1 + \cdots + a_n X_n$ is hard but finding its moment-generating function is not.

---

[2] Mathematically, one *cannot* always do this, but the function $F_Y(\cdot)$ satisfies the technical conditions under which the interchange is permissible. After all, $F_Y(\cdot)$ is not just any arbitrary function; it is a cumulative distribution function, which has many special properties. We will not elaborate on these technical details here.

The theoretical justification that we can take such a "detour" and still get the right answer is never trivial. We won't be concerned with it, but it is easy to see that dealing with (3.8) is simple in the transformed domain of moment-generating functions due to the following two simple results:

(a) The moment-generating function of the sum $X_1 + X_2 + \ldots + X_n$ is, by definition,

$$m_{X_1+X_2+\ldots+X_n}(t) = \mathbb{E}(e^{t(X_1+X_2+\ldots+X_n)}) = \mathbb{E}(e^{tX_1} \cdot e^{tX_2} \cdots e^{tX_n})$$
$$\stackrel{\star}{=} \prod_{i=1}^{n} \mathbb{E}(e^{tX_i}) = \prod_{i=1}^{n} m_{X_i}(t),$$

where the step marked by "$\star$" is due to independence. Therefore, we see that adding independent random variables simply "translates to" multiplying their moment-generating functions.

(b) The moment-generating function of a scaled random variable aX, where a is a (non-random) constant, is, by definition,

$$m_{aX}(t) = \mathbb{E}(e^{t \cdot aX}) = \mathbb{E}(e^{at \cdot X}) = m_X(at).$$

That is, scaling a random variable simply "translates to" scaling the argument of its moment-generating function.

Using these two facts (a)–(b), the moment-generating function of

$$Y = \sum_{i=1}^{n} a_i X_i$$

is simply:

$$m_Y(t) = \prod_{i=1}^{n} m_{a_i X_i}(t) = \prod_{i=1}^{n} m_{X_i}(a_i t). \tag{3.13}$$

Example 3.9. Remarkably, the central limit theorem (Theorem 1 in section 2.4), which certainly appears to be a very difficult result due to its very general applicability and surprising conclusion, actually can be proved fairly easily by (3.13).

First, notice that we can rewrite:

$$\frac{S_n - n\mu}{\sqrt{n}\sigma} = \frac{1}{\sqrt{n}}\left[\frac{X_1 - \mu}{\sigma} + \frac{X_2 - \mu}{\sigma} + \ldots + \frac{X_n - \mu}{\sigma}\right].$$

Since each $Z_i \equiv (X_i - \mu)/\sigma$ has expectation 0 and variance 1, using moment-generating functions, it suffices to prove that the moment-generating function of

$$\frac{1}{\sqrt{n}}(Z_1 + Z_2 + \ldots + Z_n), \qquad\qquad (3.14)$$

where $Z_1, Z_2, \ldots, Z_n$ are i.i.d. with $\mathbb{E}(Z_i) = 0$ and $\mathbb{V}\mathrm{ar}(Z_i) = 1$, converges to that of $N(0, 1)$.

Next, since $Z_1, Z_2, \ldots, Z_n$ all have the same distribution, they all have the same moment-generating function as well—denote it by $m(t)$. By (3.13), the moment-generating function of (3.14) is

$$\left[m\left(\frac{t}{\sqrt{n}}\right)\right]^n.$$

For large n, $t/\sqrt{n}$ is close to 0, so using a second-order Taylor approximation (see Mathematical Insert 4 below) of $m(\cdot)$ near 0 gives

$$m\left(\frac{t}{\sqrt{n}}\right) \approx m(0) + m'(0)\left(\frac{t}{\sqrt{n}}\right) + \frac{m''(0)}{2}\left(\frac{t}{\sqrt{n}}\right)^2.$$

The reason why this is useful is because, while we don't know much about $m(\cdot)$ itself,[3] we actually do know the coefficients—$m(0)$, $m'(0)$, and $m''(0)$—in the quadratic approximation above. To start, it is easy to see that

$$m(0) = \mathbb{E}(e^{tZ_i})|_{t=0} = 1.$$

But don't forget that we know $\mathbb{E}(Z_i) = 0$ and $\mathbb{V}\mathrm{ar}(Z_i) = 1$. This means

$$m'(0) = \left[\frac{d}{dt}\mathbb{E}(e^{tZ_i})\right]_{t=0} = \mathbb{E}(e^{tZ_i} \cdot Z_i)|_{t=0} = \mathbb{E}(Z_i) = 0$$

---

[3] The central limit theorem is so general that it applies to almost any distribution; thus, $m(\cdot)$ could be just about anything.

Mathematical Insert 4

Taylor approximation. Near $x = a$, a function $f(x)$ can be approximated by a polynomial,

$$f(x) \approx f(a) + f'(a)(x - a) + \frac{f''(a)}{2}(x - a)^2 + \frac{f'''(a)}{3!}(x - a)^3 + \ldots,$$

provided that the derivatives exist.

Mathematical Insert 5

A well-known limit.

$$\lim_{n \to \infty}\left(1 + \frac{x}{n}\right)^n = e^x.$$

---

and

$$m''(0) = \left[\frac{d^2}{dt^2}\mathbb{E}(e^{tZ_i})\right]_{t=0} = \mathbb{E}(e^{tZ_i} \cdot Z_i^2)|_{t=0} = \mathbb{E}(Z_i^2)$$

$$= \mathbb{V}ar(Z_i) + [\mathbb{E}(Z_i)]^2 = 1.$$

Finally, putting all of this together, we get

$$m\left(\frac{t}{\sqrt{n}}\right) \approx 1 + \frac{t^2}{2n} \quad \Rightarrow \quad \left[m\left(\frac{t}{\sqrt{n}}\right)\right]^n \approx \left[1 + \frac{t^2/2}{n}\right]^n \xrightarrow{n \to \infty} e^{t^2/2},$$

where the final step is based on a well-known limit in mathematics (see Mathematical Insert 5 above). We will leave it as an exercise for you to verify that $e^{t^2/2}$ is indeed the moment-generating function of $N(0, 1)$. □

Exercise 3.9. Let $Z \sim N(0, 1)$ and $m_Z(t)$ be its moment-generating function. Show that $m_Z(t) = e^{t^2/2}$. □

Exercise 3.10. Suppose $X_1, X_2, \ldots, X_n$ are independent random variables, each distributed as $N(\mu_i, \sigma_i^2)$ for $i = 1, 2, \ldots, n$. Use moment-generating functions to find the distribution of $S_n = a_1 X_1 + a_2 X_2 + \ldots + a_n X_n$, where $a_1, a_2, \ldots, a_n$ are (non-random) constants. [*Hint: According to Exercise 3.6 (in section 3.2.1),* *each random variable X_i here can be expressed as $X_i = \mu_i + \sigma_i Z_i$, where $Z_i \sim N(0, 1)$. First, show that the moment-generating function of each X_i is given by $m_i(t) = \exp(\mu_i t + \sigma_i^2 t^2 / 2)$.]* □

3.A.3 Formulae for expectations and variances

Sometimes, we may not need to know everything about the distribution—and it may be good enough if we can simply compute (or approximate) the expectation and variance—of $g(X_1, X_2, \ldots, X_n)$. For sums of independent random variables (3.8), these can be computed with two very simple formulae:

$$\mathbb{E}\left[\sum_{i=1}^{n} a_i X_i\right] = \sum_{i=1}^{n} a_i \mathbb{E}(X_i), \quad \mathbb{V}\mathrm{ar}\left[\sum_{i=1}^{n} a_i X_i\right] = \sum_{i=1}^{n} a_i^2 \mathbb{V}\mathrm{ar}(X_i). \quad (3.15)$$

Here, it is important for us to emphasize that the variance formula given above is *not* a general formula for sums of *any* set of random variables (whereas the expectation formula is) but a simplified version only for sums of *independent* random variables. This is all we need for this book, but we'd like to illustrate, using a simple scenario of $n = 2$, what happens if the random variables are *not* independent.

Given two random variables X and Y, we can calculate the variance of $aX + bY$, where a, b are (non-random) constants, directly using the basic variance formula (2.15):

$$
\begin{aligned}
\mathbb{V}\mathrm{ar}(aX + bY) &= \mathbb{E}[(aX + bY)^2] - [\mathbb{E}(aX + bY)]^2 \\
&= \left\{a^2\, \mathbb{E}(X^2) + b^2\, \mathbb{E}(Y^2) + 2ab\mathbb{E}(XY)\right\} - \\
&\quad \left\{a^2[\mathbb{E}(X)]^2 + b^2[\mathbb{E}(Y)]^2 + 2ab\mathbb{E}(X)\mathbb{E}(Y)\right\} \\
&= a^2\mathbb{V}\mathrm{ar}(X) + b^2\mathbb{V}\mathrm{ar}(Y) + 2ab\underbrace{[\mathbb{E}(XY) - \mathbb{E}(X)\mathbb{E}(Y)]}_{\mathrm{Cov}(X,Y)}.
\end{aligned}
$$

The term $\mathbb{E}(XY) - \mathbb{E}(X)\mathbb{E}(Y)$ is known as the *covariance* between X and Y. It is a very important concept for statistics but one that will not be covered much at all in this book. The main point here is that this extra covariance term is generally *not* equal to zero, so the variance of $aX + bY$ depends not only upon the individual variances $\mathbb{V}\mathrm{ar}(X)$ and $\mathbb{V}\mathrm{ar}(Y)$ but also upon their covariance $\mathbb{C}\mathrm{ov}(X, Y)$. Thus, to make it general, we must add all the pairwise covariance terms to the variance formula in (3.15).

If X and Y are independent, however, then

$$
\begin{aligned}
\mathbb{E}(XY) &= \iint xy f(x, y)\, dxdy \\
&= \iint xy f_X(x) f_Y(y)\, dxdy \quad \text{(due to independence)} \\
&= \int x f_X(x) \underbrace{\left[\int y f_Y(y) dy \right]}_{\mathbb{E}(Y)} dx \\
&= \mathbb{E}(Y) \underbrace{\int x f_X(x) dx}_{\mathbb{E}(X)} \\
&= \mathbb{E}(X)\mathbb{E}(Y),
\end{aligned}
$$

and the extra covariance term is zero, so the (simpler) variance formula in (3.15) holds.

Unfortunately, the converse is *not* true; that is, if $\mathbb{C}\mathrm{ov}(X, Y) = 0$, it does *not* imply that X and Y are necessarily independent. This begs the question of how the notion of "zero covariance" differs from that of independence—a very important question but one which we will skip in this book as we are not going into any depth about the concept of covariance itself.

Exercise 3.11. Let $X \sim \text{Binomial}(n, p)$. Show that $\mathbb{E}(X) = np$ and $\mathbb{V}\mathrm{ar}(X) = np(1 - p)$. [*Hint: Express X as a sum of i.i.d. Binomial$(1, p)$, or Bernoulli(p), random variables, and apply the results from Example 2.5 in section 2.3.3.*]   □

PART II

DOING STATISTICS

Synopsis: Before a model becomes truly useful, one must learn something about the unknown quantities in it (e.g. its parameters) from the data it is presumed to have generated, whether one cares about the parameters themselves or not; that's what much of statistical inference is about.

4

Overview of Statistics

In probability, we try to describe the behavior of random variables based on their distributions. For example, we cannot be certain about what might happen to random variables $X_1, \ldots, X_n$, but with their (joint) distribution, say, $f(x_1, \ldots, x_n)$, we can compute quantities such as

$$\mathbb{P}[g(X_1, \ldots, X_n) \in A] = \int_{g(x_1, \ldots, x_n) \in A} f(x_1, \ldots, x_n)\, dx_1 \ldots dx_n, \quad A \subset \mathbb{R},$$

and

$$\mathbb{E}[h(X_1, \ldots, X_n)] = \int h(x_1, \ldots, x_n) f(x_1, \ldots, x_n)\, dx_1 \ldots dx_n.$$

Thus, whereas we cannot be sure whether $g(X_1, \ldots, X_n)$ will fall into a specific set A, we can nonetheless say that such an event will happen with a certain probability; whereas we cannot be sure what the value of $h(X_1, \ldots, X_n)$ will be, we can nonetheless say that, on average, its value will be this or that.

In statistics, we are mostly concerned with the "opposite" question. We try to say something about the underlying distributions based on the random variables that they generate. (Read Chapter 1 again if it is not yet clear to you why this is what statistics is mostly about.) This process is more formally referred to as performing *statistical inference*.

Usually (and certainly within the scope of this book), the underlying distributions belong to various parametric families, and saying something about them really means saying something about their parameters. For ease of presentation, we will pretend, for the time being, that there is just one parameter—call it θ—but the same ideas apply if there are many parameters. In fact, one can simply think of θ as multidimensional, for example, $\theta = (\theta_1, \ldots, \theta_d)^\top \in \mathbb{R}^d$. We will use "$f(x_1, \ldots, x_n; \theta)$" to denote the fact that the (joint) distribution of $(X_1, \ldots, X_n)$ depends on the parameter θ; sometimes, we also say that $f(\cdot)$ is parameterized by θ.

So what kind of things can we say about the parameter θ? This differs, depending on which statistical approach one takes.

Essential Statistics for Data Science. Mu Zhu, Oxford University Press. © Mu Zhu (2023).
DOI: 10.1093/oso/9780192867735.003.0004

4.1 Frequentist approach

In the *frequentist* approach, the parameter θ is assumed to be a fixed (albeit unknown) constant. In this framework, the main thing to be said about θ is to estimate its unknown value. More specifically, we'd like to find an *estimator*,

$$\hat{\theta} = g(X_1, \ldots, X_n),$$

which is a function of the random variables generated by the underlying model, in order to estimate θ.

4.1.1 Functions of random variables

As the estimator $\hat{\theta} = g(X_1, \ldots, X_n)$ is a function of random variables coming out of the model $f(x_1, \ldots, x_n)$, it has its own distribution as well, and its distribution actually plays a rather central role in the frequentist approach. That's why we spent quite some effort earlier in section 3.2.1 describing the general principles for how to work with functions of random variables.

Why do we care so much about the distribution of estimators? When applied to a particular data set, say, $\{x_1, \ldots, x_n\}$, an *estimator* $\hat{\theta} = g(X_1, \ldots, X_n)$ would produce a numeric *estimate*, $g(x_1, \ldots, x_n)$. For example, Amy may estimate the parameter to be 0.2138 but Bob may estimate it to be 9.4305. There is no way for us to assess how good any numeric estimate is because, after all, we don't know what the correct answer should be in the first place—if we did, we wouldn't need to estimate it. What we *can* assess, however, is the statistical properties of the estimator itself, based on its distribution.

Here is an intuitive way to understand what this distribution describes. Suppose that, instead of just one data set, we actually had many data sets, all produced by the same data-generating process. Then, on each of these data sets, the estimator would produce a different numeric estimate, and together, these different numeric estimates would "trace out" a distribution, as illustrated in Figure 4.1. This is the distribution of the estimator, $\hat{\theta} = g(X_1, \ldots, X_n)$.

For students, the difference between an *estimator* and an *estimate* can seem very subtle. Perhaps a good way to distinguish them is as follows: the estimate refers to a specific answer produced by a certain procedure, but the estimator refers to the procedure itself. We cannot assess the specific answer because we don't know what the correct answer should be, but we can assess the procedure that's used to produce the answer. In some sense, this is the very essence of statistics.

$$f(x_1, \ldots, x_n; \theta) \to \begin{cases} \to \{x_1^{(1)}, \ldots, x_n^{(1)}\} & \to \hat{\theta}^{(1)} \\ \to \{x_1^{(2)}, \ldots, x_n^{(2)}\} & \to \hat{\theta}^{(2)} \\ \to \{x_1^{(3)}, \ldots, x_n^{(3)}\} & \to \hat{\theta}^{(3)} \\ \quad \vdots & \quad \vdots \\ \to \{x_1^{(B)}, \ldots, x_n^{(B)}\} & \to \hat{\theta}^{(B)} \end{cases} \Bigg\} \text{ distribution of } \hat{\theta}$$

Figure 4.1 Schematic illustration of why $\hat{\theta} = g(X_1, X_2, \ldots, X_n)$ is a random variable and has a distribution of its own.
Source: authors.

Thus, we can say $\hat{\theta}$ is a good estimator if, for example, its distribution is well concentrated near θ. This means $\hat{\theta} = g(X_1, \ldots, X_n)$ would produce a value not far from the true (but unknown) value of θ when applied to most data sets (produced by the same data-generating process).

The process described by Figure 4.1 can be simulated easily on a computer; the high-level pseudo code for creating such a simulation is given in Figure 4.2. In fact, this is the standard procedure that researchers use to demonstrate empirically that a particular procedure for estimating the parameters (and hence for learning the underlying probability model) is better than others. It is a very good exercise to try to do this on your own so you can fully appreciate this fundamental process of what statistics is about.

Exercise 4.1.   First, choose a probability model and fix its parameters (at the ground truth). Then, come up with a few different—either intuitively sensible or totally arbitrary—formulae for estimating one of the parameters. Finally, use simulation to investigate whether one particular formula is better than others.
As a specific case, consider

$$X_1 \sim N(\theta, \sigma^2) \quad \text{independently of} \quad X_2 \sim N(2\theta, \sigma^2).$$

```
for b = 1 to B
  data = generate(model)
  theta[b] = estimate(data)
end for
histogram(theta[1:B])
```

Figure 4.2 Pseudo code for simulating the process described in Figure 4.1.
Source: authors.

(Take, for example, $\theta = 5$ and $\sigma = 1$.) Use simulation to assess the following two formulae,

$$\widehat{\theta}_{est} = \frac{X_1 + 2X_2}{5} \quad \text{and} \quad \widehat{\theta}_{alt} = \frac{11X_1 + X_2}{13},$$

for estimating θ. Which formula has a higher chance of producing an answer that is closer to the true parameter value? [*Think: How did anyone come up with these strange formulae in the first place?*]   □

4.2 Bayesian approach

In the *Bayesian* approach, the parameter θ is assumed to be a random variable itself. (Since we don't know anything about it, why not treat it as random?!) But a random variable must be associated with a probability distribution, so we start by assigning one to it. This distribution, which we will denote by $\pi(\theta)$, is called the *prior distribution* of θ.

In this framework, $f(x_1, \ldots, x_n; \theta)$ is really a *conditional* distribution, more properly denoted as $f(x_1, \ldots, x_n | \theta)$, and the only thing that remains to be said about θ, now a random variable, is to find its *conditional* distribution, given $X_1, \ldots, X_n$, $\pi(\theta | x_1, \ldots, x_n)$, called the *posterior distribution* of θ. This is facilitated by the Bayes law (section 2.2.3),

$$\pi(\theta | x_1, \ldots, x_n) = \frac{f(x_1, \ldots, x_n | \theta)\pi(\theta)}{\int f(x_1, \ldots, x_n | \theta)\pi(\theta)d\theta}, \qquad (4.1)$$

which is where the name "Bayesian" comes from.

4.2.1 Compound distributions

One can now begin to appreciate why we specifically discussed the idea of compound distributions earlier in section 3.2.2. The denominator in Equation (4.1) is a compound distribution—the distribution of $X_1, \ldots, X_n$ depends on the parameter θ, which is itself a random variable with (prior) distribution $\pi(\theta)$. As a matter of fact, this compound distribution appearing in the denominator of (4.1) is the main (technological) bottleneck[1] of the Bayesian approach because the integral (or sum) that we must evaluate is predominantly intractable, especially if the dimension of θ is high—that is, if the model has many parameters, which is often the case for real applications. It is also the main reason why the Bayesian approach did not take center stage in statistics until the late 1980s, when it became widely possible to use computer algorithms to overcome this technological bottleneck, even though the Bayes law—and hence Equation (4.1) itself—dates back to as early as 1763.

4.3 Two more distributions

In Part II of this book, we discuss these different statistical approaches in more detail. Before we proceed, we would like to introduce two more distributions. Together, these two distributions are rich enough so that we are able to use them for most of our examples in Part II. This is a deliberate approach (see also Remark 3.1 in section 3.1) so that students can focus on the main ideas and principles rather than being distracted by specific tricks for dealing with different distributions.

4.3.1 Poisson distribution

A discrete random variable X is said to follow the Poisson(λ) distribution if it has probability mass function

$$f(x) = \mathbb{P}(X = x) = \frac{e^{-\lambda}\lambda^x}{x!}, \quad x \in \{0, 1, 2, \ldots\}. \tag{4.2}$$

This expression is not as intuitive as, for example, the Binomial mass function (2.16), in that one cannot simply stare at it and tell what each component means. For example, where does the exponential term $e^{-\lambda}$ come from?!

The expression (4.2) itself is deceiving. There is actually a very intimate connection between the Poisson expression (4.2) above and the Binomial

---

[1] Another bottleneck, more philosophical in nature, is the choice of the prior distribution, $\pi(\theta)$. We will say more about both of these bottlenecks in Chapter 6, when we discuss the Bayesian approach in more detail.

expression (2.16) earlier. In particular, by defining a fixed constant $\lambda = np$ and letting n go to infinity in the Binomial expression, we obtain

$$\binom{n}{x} p^x (1 - p)^{n-x} = \frac{n!}{x!(n-x)!} \left[\frac{\lambda}{n}\right]^x \left[1 - \frac{\lambda}{n}\right]^{n-x}$$

$$= \frac{\lambda^x}{x!} \cdot \underbrace{\frac{n(n-1)(n-2)\ldots(n-x+1)}{n^x}}_{\downarrow \atop 1} \cdot \underbrace{\left[1 - \frac{\lambda}{n}\right]^n}_{\downarrow \atop e^{-\lambda}} \cdot \underbrace{\left[1 - \frac{\lambda}{n}\right]^{-x}}_{\downarrow \atop 1}$$

$$\longrightarrow \frac{e^{-\lambda} \lambda^x}{x!},$$

that is, the Poisson expression!

This is not just a piece of "fancy" mathematics for anyone to show off with. As $n \to \infty$, $p = \lambda/n$ must necessarily go down to 0 if λ is a fixed constant. What does this mean for the Binomial distribution? It means we are conducting many, many experiments ($n \uparrow \infty$), but each experiment has a very, very small success probability ($p \downarrow 0$). In other words, this piece of mathematics actually explains why the Poisson distribution can be, and often is, used to model the occurrence of rare events in a large population, such as the number of serious traffic accidents, incidences of a rare infection, and so on.

For $X \sim \text{Poisson}(\lambda)$, we have

$$\mathbb{E}(X) = \lambda \quad \text{and} \quad \mathbb{V}\text{ar}(X) = \lambda.$$

Knowing that the Poisson is a limit of the Binomial allows us to see these results immediately. For $X \sim \text{Binomial}(n, p)$, we have $\mathbb{E}(X) = np$ and $\mathbb{V}\text{ar}(X) = np(1 - p)$ (Exercise 3.11 in section 3.A.3). But in the Poisson limit, we define $\lambda = np$ and let n go to infinity. So $\mathbb{E}(X) = \lambda$ and $\mathbb{V}\text{ar}(X) = \lambda(1 - \lambda/n) \to \lambda$. A direct derivation of these results requires a few special tricks; they are given in Appendix 4.A as the tricks themselves—though worth learning—are not crucial for continuing with this book.

Exercise 4.2. Let $X \sim \text{Binomial}(n, p)$. If the parameter n is also random, then X has a compound distribution. Find $\mathbb{P}(X = x)$ if $n \sim \text{Poisson}(\lambda)$. [*Note: For instance, X may be the number of sales at a certain car dealership today, and n may be the number of customers. See also Example 3.8 in section 3.2.2.*]   □

4.3.2 Gamma distribution

A continuous random variable X is said to follow the Gamma(α, β) distribution if it has probability density function

$$f(x) = \frac{\beta^\alpha}{\Gamma(\alpha)}x^{\alpha-1}e^{-\beta x}, \quad x > 0.$$

With two free parameters, $\alpha, \beta > 0$, this is a reasonably flexible model for many non-negative, continuously valued random quantities. For example, by varying these two parameters, the function $f(x)$ can take on many different shapes, thus providing a reasonably good fit for describing many random phenomena. As such, it is widely used, even though there is *not* always a compelling physical interpretation of what the function actually represents.

One well-known case where the Gamma distribution *can* be derived physically is if we have a *Poisson process*[2] with rate parameter λ. Then, the amount of time that one must wait till the n-th occurrence is a random variable, and its distribution can be derived to be Gamma(n, λ); see Appendix 4.B.

Below, we derive the expectation of the Gamma distribution. In doing so, we highlight a very important practical "technique" for integration that will become useful later in this book. By definition,

$$
\begin{aligned}
\mathbb{E}(X) &= \int_0^\infty x \cdot \frac{\beta^\alpha}{\Gamma(\alpha)}x^{\alpha-1}e^{-\beta x}\,dx \\
&= \frac{\beta^\alpha}{\Gamma(\alpha)}\int_0^\infty x^{(\alpha+1)-1}e^{-\beta x}\,dx \\
&= \frac{\beta^\alpha}{\Gamma(\alpha)}\cdot\frac{\Gamma(\alpha+1)}{\beta^{\alpha+1}}\int_0^\infty \underbrace{\frac{\beta^{\alpha+1}}{\Gamma(\alpha+1)}x^{(\alpha+1)-1}e^{-\beta x}}_{\text{density of Gamma}(\alpha+1,\,\beta)}\,dx \\
&= \frac{\beta^\alpha}{\Gamma(\alpha)}\cdot\frac{\Gamma(\alpha+1)}{\beta^{\alpha+1}}\cdot 1 \\
&= \frac{\alpha}{\beta},
\end{aligned}
$$

where the last step is based on the recursive property of the Gamma function (see Mathematical Insert 3 in section 3.2.1).

---

[2] The Poisson process is a special *stochastic process* that counts the number of times a certain event has happened as time progresses, for example, the number of system intrusions after the system has been operating for one hour, two hours and so on. It is a more advanced topic in probability theory. We only mention this in passing.

As illustrated, the aforementioned "technique" is to exploit the fact that any density function must integrate to one. This description or summary of the "technique" makes it sound so trivial and obvious that one is hardly justified to even use the name "technique", but it is tremendously helpful and will make life much easier if one truly masters it.

Remark 4.1. In the special case of $\alpha = 1$, the Gamma$(1, \beta)$ distribution is more commonly known as the Exponential(β) distribution. □

Exercise 4.3. Let $X \sim$ Gamma(α, β). Find $\mathbb{V}\text{ar}(X)$. □

Exercise 4.4. Let $X_1, \ldots, X_n$ be independent random variables, each distributed as Gamma(α_i, β), respectively for $i = 1, 2, \ldots, n$.

(a) Let $S_n = X_1 + \ldots + X_n$. Show that $S_n \sim$ Gamma$(\sum_{i=1}^{n} \alpha_i, \beta)$. [*Hint: Use the convolution formula (3.11) to show this for $n = 2$ and argue by induction. Alternatively, find the moment-generating function of X_i and proceed using fact (a) of section 3.A.2.*]

(b) Verify that the Gamma$(1/2, 1/2)$ distribution is the same as the $\chi^2_{(1)}$ distribution given in Exercise 3.5 (section 3.2.1).

(c) Suppose $X_1, X_2, \ldots, X_n \overset{iid}{\sim} N(0, 1)$ and $Y_n = X_1^2 + X_2^2 + \cdots + X_n^2$. Show that $Y_n \sim$ Gamma$(n/2, 1/2)$.

[*Note: The Gamma$(n/2, 1/2)$ distribution is also known as the "chi-squared distribution with n degrees of freedom," often denoted as $\chi^2_{(n)}$.*] □

Exercise 4.5. Bond, James Bond, has just been ordered by M to destroy a nuclear facility in North Korea. Let N be the number of attempts he must make until he succeeds. Clearly, N is a random variable. Suppose it has probability mass function:

$$f_N(n) = (1 - p)^{n-1} p, \quad n \in \{1, 2, 3, \ldots\}.$$

Meanwhile, Q is waiting for M to give him more money to start building his next gadget. Let T be the amount of time Q has to wait. In fact, T is dependent upon how quickly Bond can succeed in his North Korean mission. Specifically, conditional on $N = n$, $T|(N = n) \sim$ Gamma(n, λ) with

$$f_{T|N}(t|n) = \frac{\lambda^n}{\Gamma(n)} t^{n-1} e^{-\lambda t}, \quad t > 0.$$

(a) Find $f_T(t)$, the marginal distribution of T.

(b) By a given time, say, t_0, what is the probability that Q still hasn't got his gadget money?

(c) Find $f_{N|T}(n|t)$, the conditional distribution N given T. [*Think: Do you recognize this distribution? How would you characterize it?*]

(d) For a price of $£100 \times (p + 0.01)$, employees of MI6 can buy a special bond (pun intended) from M. After James returns, the bond will be redeemable for its face value of $£100$ only if James turns out to have succeeded on his first attempt ($N = 1$) but will be worthless otherwise ($N > 1$). "Clearly a bad deal for us", thought Q. [*Think: Why?*]. But he quickly changed his mind when he got his gadget money at time t_0. What condition must t_0 satisfy for Q to change his mind?

[*Hint: Part (d) has to do with pricing risky investments. Usually, the expected pay-off is considered to be the fair price. But, of course, this expectation changes as relevant information becomes available.*] □

Appendix 4.A Expectation and variance of the Poisson

Even though the notion of $\mathbb{E}(X)$ and $\mathbb{V}\mathrm{ar}(X)$ may be clear, it does not mean computing them is always easy; in fact, they can often be quite hard. For special distributions, there are often some special tricks that can be used. For the Poisson, which we introduced earlier in this chapter, its expectation is:

$$
\begin{aligned}
\mathbb{E}(X) &= \sum_{x=0}^{\infty} x \cdot \frac{e^{-\lambda}\lambda^x}{x!} \\
&= \sum_{x=1}^{\infty} x \cdot \frac{e^{-\lambda}\lambda^x}{x!} \quad \text{(because the summand is 0 when } x = 0) \\
&= e^{-\lambda} \sum_{x=1}^{\infty} \frac{\lambda^x}{(x-1)!} \\
&= e^{-\lambda}\lambda \sum_{x=1}^{\infty} \frac{\lambda^{x-1}}{(x-1)!} \\
&= e^{-\lambda}\lambda \underbrace{\sum_{y=0}^{\infty} \frac{\lambda^y}{y!}}_{e^\lambda} \quad \text{(by letting } y = x - 1) \\
&= \lambda.
\end{aligned}
$$

We see that a key "trick" here is the cancellation of the leading term x from $x!$, giving us $(x - 1)!$.

The variance of the Poisson can be computed by employing the same "trick", but its application is more subtle. Specifically, we know that $\mathrm{Var}(X) = \mathbb{E}(X^2) - [\mathbb{E}(X)]^2$, but we easily get stuck when trying to compute $\mathbb{E}(X^2)$ directly. (Try it yourself!) So what shall we do? The answer is that we compute $\mathbb{E}[X(X - 1)]$ instead! Why? Because this allows us to apply the same "trick"—cancelling out the leading terms $x(x - 1)$ from $x!$ to obtain $(x - 2)!$, but, at the same time, it also allows us to obtain $\mathbb{E}(X^2)$ *indirectly*, from the fact that $\mathbb{E}[X(X - 1)] = \mathbb{E}(X^2 - X) = \mathbb{E}(X^2) - \mathbb{E}(X)$, but we have already computed $\mathbb{E}(X) = \lambda$. We will leave it as an exercise for you to complete these steps.

Exercise 4.6. Let $X \sim$ Poisson(λ). Show that $\mathrm{Var}(X) = \lambda$. [*Hint: Compute* $\mathbb{E}[X(X - 1)]$ *first.*] ☐

Exercise 4.7. Your company is running an incentive program to promote workplace safety. Each month, your unit will be given an incentive amount of $1/(X + 1)$, where X is the number of workplace accidents your unit has had in that month. (Hence, if you have no accidents, you will get the full incentive; if you have one accident, you will get half of the incentive; if you have two accidents, you will only get one-third of the incentive; and so on.) However, the company is relying on a mysterious offshore connection to fund this program, and your unit will most probably not receive the money immediately. Therefore, at the end of the month, the discounted value of your incentive is equal to

$$\left[\frac{1}{X + 1}\right] e^{-rT},$$

where T is the extra amount of time you have to wait for the money to arrive and r is the risk-free interest rate (assumed to be a fixed constant). Assume that

$$X \sim \text{Poisson}(\lambda) \quad \text{and} \quad T \sim \text{Gamma}(\alpha, \beta).$$

(a) On average, how much do you expect the discounted value of your incentive to be, assuming that the two random quantities X and T are independent?

(b) How reasonable is the independence assumption in part (a)?

[*Fact: If X and Y are independent, then* $\mathbb{E}[g(X)h(Y)] = \mathbb{E}[g(X)]\mathbb{E}[h(Y)]$. *Furthermore, if $X \sim f(x)$, then* $\mathbb{E}[g(X)] = \int g(x) f(x) dx$ *if X is continuous, and* $\mathbb{E}[g(X)] = \sum g(x) f(x)$ *if X is discrete.*] ☐

Appendix 4.B Waiting time in Poisson process

For a Poisson process with rate parameter λ, the number of events to occur by time t is a Poisson random variable with parameter λt. This is a defining property of the Poisson process.

Since we haven't formally introduced the Poisson process, or any stochastic process for that matter, we won't say much about this property itself. The main objective of this section is to show how one can derive the Gamma distribution from this particular property of the Poisson process.

Let T_n denote the amount of time one must wait till the n-th event. The key observation here is that one has to wait for at least t units of time (i.e. $T_n > t$) if, and only if, there have been at the most $n - 1$ events by time t because if there had been $\geq n$ events, then one wouldn't have had to wait for that long. This observation gives us enough information to compute the cumulative distribution function of T_n as

$$
\begin{aligned}
F(t) &= \mathbb{P}(T_n \leq t) \\
&= 1 - \mathbb{P}(T_n > t) \\
&= 1 - \mathbb{P}(\text{at the most } n - 1 \text{ events by time } t) \\
&= 1 - \mathbb{P}(0 \text{ event or 1 event or } \ldots \text{ or } n - 1 \text{ events by time } t) \\
&= 1 - \sum_{i=0}^{n-1} \frac{e^{-\lambda t}(\lambda t)^i}{i!},
\end{aligned}
$$

which we can then differentiate and derive the density function of T_n to be

$$
\begin{aligned}
f(t) &= \frac{d}{dt} F(t) \\
&= (-1) \sum_{i=0}^{n-1} \frac{[e^{-\lambda t}(-\lambda)](\lambda t)^i + (e^{-\lambda t})[i(\lambda t)^{i-1}\lambda]}{i!} \\
&= (-1) \left[\sum_{i=0}^{n-1} \frac{[e^{-\lambda t}(-\lambda)](\lambda t)^i}{i!} + \underbrace{\sum_{i=0}^{n-1} \frac{(e^{-\lambda t})[i(\lambda t)^{i-1}\lambda]}{i!}}_{0 + \sum_{i=1}^{n-1} \frac{(e^{-\lambda t})[(\lambda t)^{i-1}\lambda]}{(i-1)!}} \right] \\
&= \sum_{i=0}^{n-1} \frac{[e^{-\lambda t}(\lambda)](\lambda t)^i}{i!} - \sum_{j=0}^{n-2} \frac{(e^{-\lambda t})[(\lambda t)^j \lambda]}{j!} \quad \text{(by letting } j = i - 1)
\end{aligned}
$$

$$= \frac{[e^{-\lambda t}(\lambda)](\lambda t)^{n-1}}{(n-1)!}$$

$$= \frac{\lambda^n}{\Gamma(n)} t^{n-1} e^{-\lambda t},$$

which is the Gamma(n, λ) density function. This derivation also demonstrates the importance of the cumulative distribution function (CDF) for continuous random variables (see section 2.3.1).

5

Frequentist Approach

This chapter focuses on the frequentist approach to "saying something about" the model parameters—specifically, how to estimate their values and why some ways of estimating them may be better than others.

5.1 Maximum likelihood estimation

For some probability models, especially simple ones, a good intuitive understanding may be more than enough for us to come up with ways to estimate its parameters.

For example, given $X \sim \text{Binomial}(n, p)$, how should we estimate the parameter p when n is also known? This model is most intuitive. If I tried to do something (e.g. throwing a dart) independently for a total of n times and obtained a total of x successes (e.g. hitting the bullseye), then an "obvious" estimate of p—my success probability at each trial—would be x/n, the fraction of time that I succeeded. This means we can use $\hat{p} = X/n$ as an *estimator* of p.

But what if the probability model is rather complex and we lack the kind of good intuition to do what we just did above? We need a certain "recipe" that we can follow regardless of how complex the model is. A widely used "recipe" is *maximum likelihood estimation.*

Let $f(x_1, x_2, \ldots, x_n; \theta)$ be the joint probability mass or density function of random variables $X_1, X_2, \ldots, X_n$, with parameter θ. When viewed as a function of θ itself, this function is called the *likelihood function* of θ:

$$L(\theta) \quad = \quad f(x_1, x_2, \ldots, x_n; \theta). \tag{5.1}$$

The maximum likelihood estimate (MLE) of θ is simply

$$\hat{\theta}_{mle} \quad = \quad \underset{\theta}{\arg\max} \quad L(\theta). \tag{5.2}$$

Essential Statistics for Data Science. Mu Zhu, Oxford University Press. © Mu Zhu (2023).
DOI: 10.1093/oso/9780192867735.003.0005

Hence, the intuitive idea here is to estimate the unknown parameter of the probability model by maximizing the joint distribution of the random variables generated by the model itself—that is, to make what comes out of the model also the most probable according to the model.

The model $X \sim$ Binomial(n, p) again allows us to see more concretely what this means. Suppose that I threw $n = 10$ darts independently. Let

$$E_7 = \{X = 7\}$$

denote the event that I succeeded at hitting the bullseye for a total of seven times, and the objective is to estimate my success probability at each trial, p. Following the maximum likelihood principle, we see that

$$\text{if} \quad p = 0.1, \qquad \mathbb{P}(E_7) = \binom{10}{7}(0.1)^7(1 - 0.1)^{10-7} \approx 8.75 \times 10^{-6};$$

$$\text{if} \quad p = 0.2, \qquad \mathbb{P}(E_7) = \binom{10}{7}(0.2)^7(1 - 0.2)^{10-7} \approx 7.86 \times 10^{-4};$$

$$\vdots \qquad\qquad\qquad \vdots$$

and so on. After trying all possible values of p (see Figure 5.1), it turns out that the event E_7 is most probable—with a probability of approximately 0.267—when $p = 0.7$. (See also Exercise 5.1 in section 5.1.1.) Thus, the MLE of p based on E_7 would be $\hat{p}_{mle} = 0.7$, and it should come as no surprise to us that this answer coincides with the intuitive approach, described at the beginning of this section, of estimating p based on the fraction of time that I succeeded.

Remark 5.1. The intuitive approach mentioned so far can be formally presented as a systematic "recipe", too. The basic principle is as follows. Suppose X is a random quantity following a certain distribution. Then, for any given function $g(\cdot)$, the value of $\mathbb{E}[g(X)]$ can be estimated (or numerically approximated) by

$$\mathbb{E}[g(X)] \approx \frac{1}{B} \sum_{b=1}^{B} g(x_b), \qquad (5.3)$$

if $x_1, x_2, \ldots, x_B$ are multiple realizations (or samples) from the *same* distribution. The left-hand side of Equation (5.3) is the theoretical mean of the random quantity, $g(X)$; the right-hand side is the empirical average of its realizations, $\{g(x_1), g(x_2), \ldots, g(x_B)\}$. This is sometimes referred

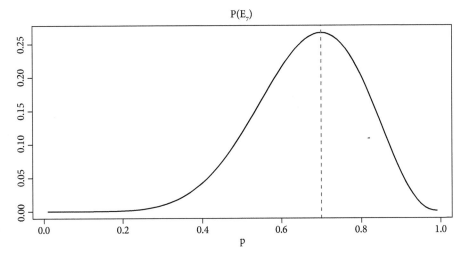

Figure 5.1 $\mathbb{P}(E_7)$ as a function of p.

Source: authors.

to as the substitution principle, the Monte Carlo method, or the method of moments—depending on context; it is an intuitive but quintessential statistical principle.

In the foregoing example, there is just $B = 1$ realization, x, from the Binomial(n, p) distribution. Then, taking $g(t) = t$ to be the identity function in Equation (5.3) gives $np \approx x$; thus, the substitution principle tells us the unknown parameter p should be approximately x/n. □

5.1.1 Random variables that are i.i.d.

If the random variables $X_1, X_2, \ldots, X_n$ are *independent and identically distributed* (i.i.d.), then the likelihood function (5.1) is simply the product of their respective marginal distributions, that is,

$$L(\theta) = \prod_{i=1}^{n} f(x_i; \theta). \tag{5.4}$$

Taking the logarithm gives us the so-called *log-likelihood function*,

$$\ell(\theta) = \sum_{i=1}^{n} \log[f(x_i; \theta)]. \tag{5.5}$$

Clearly, the maximizer of $\ell(\theta)$ and that of $L(\theta)$ are the same as the logarithmic transformation is monotonic, but algebraically it is more convenient to maximize $\ell(\theta)$ than it is to maximize $L(\theta)$—because it is much less cumbersome to differentiate a sum, for example,

$$\frac{d}{d\theta}(A + B) = \frac{d}{d\theta}A + \frac{d}{d\theta}B,$$

than it is to differentiate a product, for example,

$$\frac{d}{d\theta}(A \times B) = A \times \frac{d}{d\theta}B + B \times \frac{d}{d\theta}A.$$

That is why in practice we often work with the log-likelihood function $\ell(\theta)$ rather than the likelihood function $L(\theta)$ itself.

Example 5.1. Suppose $X_1, X_2, \ldots, X_n \overset{iid}{\sim}$ Poisson(λ). Then, the likelihood function is

$$L(\lambda) = \prod_{i=1}^{n} \frac{e^{-\lambda}(\lambda)^{x_i}}{x_i!}$$

and the log-likelihood function is

$$\ell(\lambda) = \sum_{i=1}^{n} -\lambda + x_i \log(\lambda) - \log(x_i!).$$

The first-order condition is

$$\ell'(\lambda) = 0 \quad \Rightarrow \quad \sum_{i=1}^{n}(-1) + \frac{x_i}{\lambda} = 0.$$

Solving it gives the MLE

$$\hat{\lambda}_{mle} = \frac{1}{n}\sum_{i=1}^{n} x_i = \bar{x}. \tag{5.6}$$

In simple scenarios such as this, it is often trivial to check the second-order condition. Here,

$$\ell''(\lambda) = (-1)\sum_{i=1}^{n} \frac{x_i}{\lambda^2} \le 0$$

since all x_is must be non-negative integers for the Poisson. In fact, the inequality is strict unless all of $x_1,\ldots,x_n$ are zero—surely a degenerate case where we do not have enough information to estimate the parameter λ.

This ensures that the stationary point we have found by solving the first-order condition is indeed a maximizer of $\ell(\lambda)$.

Equation (5.6) expresses the MLE $\hat{\lambda} = g(x_1,\ldots,x_n)$ as a function of $x_1,\ldots,x_n$, that is, a numeric *estimate*. The corresponding *estimator* is simply $\hat{\lambda} = g(X_1,\ldots,X_n)$, or $\hat{\lambda}_{mle} = \bar{X}$ in this case.   □

Remark 5.2. Readers will have noticed that, in the preceding example, we have used the same notation, $\hat{\lambda}_{mle}$, to refer to both the numeric *estimate*, $\bar{x}$, and the corresponding *estimator*, $\bar{X}$. Conceptually, this can be rather confusing, but it is indeed the convention of our discipline to do so as it is almost never ambiguous in specific contexts whether the same notation is referring to an *estimate* or the *estimator* itself.   □

Remark 5.3. For one-dimensional problems, we only need to worry about whether the stationary point is a minimum or a maximum. For multidimensional problems, however, we also need to worry about whether it is a *saddle point* as well. There is some evidence [6] to suggest that, for highly complex models (such as a deep neural network), saddle points can be a serious headache.   □

Exercise 5.1. Let $X \sim$ Binomial(n, p). Show that $\hat{p}_{mle} = X/n$.   □

Exercise 5.2. Let $X_1, X_2, \ldots, X_n \overset{iid}{\sim} N(\mu, \sigma^2)$. Show that

$$\hat{\mu}_{mle} = \frac{1}{n}\sum_{i=1}^{n} X_i = \bar{X} \quad \text{and} \quad \hat{\sigma}^2_{mle} = \frac{1}{n}\sum_{i=1}^{n}(X_i - \bar{X})^2.$$

[Note: Perhaps strangely, $\hat{\sigma}^2_{mle}$ has not traditionally been the default estimator that statisticians use in practice to estimate σ^2. Exercise 5.7, later in this chapter, will shed some light on this curious phenomenon.]   □

Remark 5.4. For the multivariate normal distribution, $X_1, X_2, \ldots, X_n \overset{iid}{\sim} N(\mu, \Sigma)$, $X_i \in \mathbb{R}^d$, the MLEs of the parameters are direct analogues of the univariate case above:

$$\hat{\mu} = \frac{1}{n}\sum_{i=1}^{n} X_i = \bar{X} \quad \text{and} \quad \hat{\Sigma} = \frac{1}{n}\sum_{i=1}^{n}(X_i - \bar{X})(X_i - \bar{X})^{\top}.$$

Note that each $(X_i - \bar{X})(X_i - \bar{X})^{\top}$ is $d \times d$.   □

5.1.2 Problems with covariates

We will now look at a much more interesting example. We will follow this example for quite a while in this chapter. There will be some interesting twists along the way, which will allow us to motivate and introduce some advanced techniques.

Example 5.2. Suppose $X_1, X_2, \ldots, X_n$ are independent random variables, each distributed as Poisson(λ_i), where $\lambda_i = h(v_i)$, for $i = 1, 2, \ldots, n$.

Compared with all the simple models we have looked at so far, this model is somewhat different—and hence much more interesting—in that each X_i has a different parameter that depends on another variable specific to i, namely, v_i in this case. Quantities like v_i are called *covariates* in statistics; they are usually called *predictors* in machine learning. Most useful models in practice would have covariates like we do here.

For example, X_i may be the number of accidents driver i has had in the past five years. In section 4.3.1, we explained why the Poisson distribution may be a reasonable probability model to describe the number of accidents. But clearly it is not a good idea to assume that every driver could be described by the same model parameter! For example, it surely is more likely for someone who simply drives a lot to have more accidents. Thus, v_i may represent the total mileage of driver i in the same period, and it makes sense that v_i should influence the model parameter λ_i that ultimately governs the behavior of X_i.

We are interested in how we can estimate the function $h(\cdot)$ based on $\{(v_i, X_i)\}_{i=1}^{n}$. In order to do so, we must impose some restrictions on $h(\cdot)$, the simplest of which is to postulate that $h(\cdot)$ is linear (e.g. $h(v_i) = \theta v_i$), so that the problem of estimating $h(\cdot)$ is reduced to that of simply estimating the parameter θ. (See Remark 5.5 below about the possibility of imposing much looser restrictions on h.)

Again, the method of maximum likelihood allows us to proceed without much thought. Simply write down the likelihood function,

$$L(\theta) = \prod_{i=1}^{n} \frac{e^{-\theta v_i}(\theta v_i)^{x_i}}{x_i!},$$

and then the log-likelihood,

$$\ell(\theta) = \sum_{i=1}^{n} -\theta v_i + x_i \log(\theta v_i) - \log(x_i!). \tag{5.7}$$

The first-order condition then gives

$$\ell'(\theta) = 0 \quad \Rightarrow \quad -\sum_{i=1}^{n} v_i + \sum_{i=1}^{n} x_i \frac{1}{\theta v_i} v_i = 0,$$

so the estimator based on maximum likelihood is

$$\hat{\theta}_{mle} = \frac{\sum_{i=1}^{n} X_i}{\sum_{i=1}^{n} v_i}. \tag{5.8}$$

Looking at it in hindsight, this estimator also makes a lot of intuitive sense. The denominator is simply the total mileage and the numerator the total number of accidents, across all n drivers. ☐

Remark 5.5. The model certainly can be generalized by choosing $h(\cdot)$ to be some other, more flexible, function. For instance, it is possible to model the function $h(\cdot)$ nonparametrically, requiring only that it must satisfy the condition:

$$\int [h''(t)]^2 \, dt < s,$$

for some $s > 0$, which is a restriction on its smoothness. Needless to say, the details will require us to know more mathematics than we are aiming at in this book. ☐

Given pairs of data, $\{(v_1, x_1), \ldots, (v_n, x_n)\}$, it is very common to postulate that the data-generating process is

$$X_i \sim N(\alpha + \beta v_i, \sigma^2) \quad \text{independently for } i = 1, 2, \ldots, n. \tag{5.9}$$

This is called a *simple linear regression* model. We will leave it as an exercise (Exercise 5.3) for you to work out the MLEs of the parameters, but why is it called a "regression" model? See Fun Box 1 below for an explanation of where the name "regression" came from.

Exercise 5.3. Given pairs of data, $\{(v_1, x_1), \ldots, (v_n, x_n)\}$, we postulate that the data-generating process is described by Equation (5.9). Show that, under such a model, the maximum likelihood estimates of α and β are given by

$$\hat{\alpha} = \bar{x} - \hat{\beta}\bar{v} \quad \text{and} \quad \hat{\beta} = \frac{\sum_{i=1}^{n} (v_i - \bar{v})(x_i - \bar{x})}{\sum_{i=1}^{n} (v_i - \bar{v})^2},$$

respectively. [*Note: One can also derive the MLE of σ^2, but we will focus on just α and β here.*] ☐

Fun Box 1

Regression to the mean. When analyzing pairs of data using the simple linear regression model [Equation (5.9)], it is common to standardize each x_i prior to the analysis by the sample mean and sample standard deviation, with

$$x_i \leftarrow \frac{x_i - \bar{x}}{s_x}, \quad \text{where} \quad s_x = \sqrt{\frac{1}{n-1} \sum_{i=1}^{n} (x_i - \bar{x})^2};$$

and likewise for $v_i \leftarrow (v_i - \bar{v})/s_y$. This means both quantities are measured in terms of standard deviation units above or below their respective averages. Then, using results from Exercise 5.3 (section 5.1.2) and the Cauchy–Schwarz inequality (Mathematical Insert 6 below), one can show that the MLEs of α and β will satisfy the following conditions:

$$\hat{\alpha} = 0 \quad \text{and} \quad |\hat{\beta}| \leq 1.$$

(Try it.) In a famous study [7], it was observed that very tall (or short) fathers—say, those with a height of $v = \pm 2$ standard deviation units away from the average father—had sons who were, *on average*, not as tall (or short) as their fathers; for example, their average height was only about $0.5(v) = (0.5)(\pm 2) = \pm 1$ standard deviation units away from the average son. Since then, many such phenomena have been observed; for example, students who did exceedingly well (or badly) on the first test tend to do *not* as well (or as badly) on the second one. Sir Francis Galton used the phrase, "regression towards mediocrity" to describe such phenomena, which is where the name "regression" originally came from. Today, the phrase, "regression to the mean" is more often used.

Mathematical Insert 6

Cauchy–Schwarz inequality.

$$\left(\sum_{i=1}^{n} a_i^2 \right) \left(\sum_{i=1}^{n} b_i^2 \right) \geq \left(\sum_{i=1}^{n} a_i b_i \right)^2.$$

If we define two vectors, $a, b \in \mathbb{R}^n$, then the expression above merely says $\|a\|^2 \|b\|^2 \geq \langle a, b \rangle^2$; that is, the product of their squared *norms* is at least as large as the square of their *inner product*. In fact, this is the more general statement of the Cauchy–Schwarz inequality, and it holds in any space with a norm and an inner product, for example,

$$\left(\int f^2(x)dx\right)\left(\int g^2(x)dx\right) \geq \left(\int f(x)g(x)dx\right)^2.$$

That is why it is one of the most important and useful inequalities in mathematics.

5.2 Statistical properties of estimators

As we have already alluded to in section 4.1, when assessing estimators, it is their statistical properties that we are concerned with, not the numeric values they produce on specific data sets, for there is no way to assess the latter. And their statistical properties are, of course, encapsulated in their distributions.

Example 5.3. Recall Example 5.1 (section 5.1.1). By the central limit theorem (Theorem 1 in section 2.4), the distribution of $\widehat{\lambda}_{mle} = \bar{X}$ is approximately $N(\lambda, \lambda/n)$. In fact, the only part that is "approximate" in the preceding statement is the part about normality; the expectation and variance of $\bar{X}$ are not approximate but exact—this can be checked easily by letting $a_i = 1/n$ in Equation (3.15) in section 3.A.3. (Do it.)

Therefore, we see that $\mathbb{E}(\widehat{\lambda}_{mle}) = \lambda$, which means that, on average, the estimator $\widehat{\lambda}_{mle}$ will give the correct answer. Moreover, we see that $\mathbb{Var}(\widehat{\lambda}_{mle}) \to 0$ as $n \to \infty$, which means that, as the sample size grows, we can be quite sure that the estimator $\widehat{\lambda}_{mle}$ will give an answer that is not too far from the correct one, namely λ. (See also Exercise 5.4 below.)

Notice how we are able to say all these things about the estimator $\widehat{\lambda}_{mle}$ without ever knowing what the true value of λ actually is! □

Exercise 5.4. Continue with Example 5.3 above. Apply Chebyshev's inequality (see Exercise 2.6 in section 2.3.2) to show that

$$\mathbb{P}(|\widehat{\lambda}_{mle} - \lambda| > \varepsilon) \to 0 \quad \text{as} \quad n \to \infty$$

for any given $\varepsilon > 0$. [*Note: An estimator $\widehat{\theta}_n$ (of θ) is said to be "consistent" if $\mathbb{P}(|\widehat{\theta}_n - \theta| > \varepsilon) \to 0$ as $n \to \infty$.*] □

Just as the expectation and the variance are two important summaries of any distribution (section 2.3.2), so $\mathbb{E}(\widehat{\theta})$ and $\mathbb{V}\mathrm{ar}(\widehat{\theta})$ are two important statistical properties of any estimator $\widehat{\theta}$—one quantifies its *bias* and the other its *efficiency*.

Definition 5 (Unbiasedness). *An estimator $\widehat{\theta}$ is said to be unbiased if $\mathbb{E}(\widehat{\theta}) = \theta$. Consequently, the bias of $\widehat{\theta}$ is defined to be $\mathbb{B}\mathrm{ias}(\widehat{\theta}) = \mathbb{E}(\widehat{\theta}) - \theta$.* □

Definition 6 (Efficiency). *If $\widehat{\theta}_1$ and $\widehat{\theta}_2$ are both unbiased estimators of θ, then we say $\widehat{\theta}_1$ is more efficient than $\widehat{\theta}_2$ if $\mathbb{V}\mathrm{ar}(\widehat{\theta}_1) < \mathbb{V}\mathrm{ar}(\widehat{\theta}_2)$.* □

Definition 6 above makes it clear that we can only compare the relative efficiency of two *unbiased* estimators. This makes sense—we must compare apples with apples. But is there a way for us to compare apples with oranges, for example, perhaps by their overall nutritional value? How can we compare estimators more generally? One way to do so is to look at an estimator's overall *mean squared error* (MSE),

$$
\begin{aligned}
\mathrm{MSE}(\widehat{\theta}) &\equiv \mathbb{E}[(\widehat{\theta} - \theta)^2] & (5.10) \\
&= \mathbb{E}\{[\widehat{\theta} - \mathbb{E}(\widehat{\theta}) + \mathbb{E}(\widehat{\theta}) - \theta]^2\} \\
&= \mathbb{E}\{[\widehat{\theta} - \mathbb{E}(\widehat{\theta})]^2\} + [\mathbb{E}(\widehat{\theta}) - \theta]^2 + 2\underbrace{\mathbb{E}\{[\widehat{\theta} - \mathbb{E}(\widehat{\theta})][\mathbb{E}(\widehat{\theta}) - \theta]\}}_{\overset{\star}{=}0} \\
&= \mathbb{V}\mathrm{ar}(\widehat{\theta}) + [\mathbb{B}\mathrm{ias}(\widehat{\theta})]^2, & (5.11)
\end{aligned}
$$

which, as the foregoing steps show, can be decomposed into a bias component and a variance component. This bias–variance decomposition of the MSE, Equation (5.11), is fundamental for statistics.

Exercise 5.5. Complete the derivation from Equation (5.10) to Equation (5.11) by proving the step marked by "$\star$" above. □

Example 5.4. Recall Example 5.2 (section 5.1.2), where $\widehat{\theta}_{mle}$, given by Equation (5.8), had an intuitive interpretation as an overall accident-to-mileage ratio, that is,

$$
\widehat{\theta}_{mle} = \frac{\text{(total number of accidents across all } n \text{ divers)}}{\text{(total mileage across all } n \text{ drivers)}}.
$$

What if we took a slightly different point of view? Instead of one data set of size n, what if we proceeded as if we had n separate "data sets", each of size one? Then, on each of these "data sets", the same accident-to-mileage estimator would be equal to X_i/v_i, for $i = 1, 2, \ldots, n$. We could then average these n "individual MLEs" to obtain an overall estimator,

$$\hat{\theta}_{alt} = \frac{1}{n} \sum_{i=1}^{n} \frac{X_i}{v_i}. \tag{5.12}$$

Notice that this is a little different from what we obtained earlier in Equation (5.8). This approach is not at all unreasonable in itself, and we now have two competing estimators, $\hat{\theta}_{mle}$ and $\hat{\theta}_{alt}$, but which one is superior?

In this case, both estimators are linear functions of $X_1, \ldots, X_n$—just two *different* linear functions. Using the formulae contained in Equation (3.15) in section 3.A.3, we can first compute that

$$\mathbb{E}(\hat{\theta}_{mle}) = \frac{\sum_{i=1}^{n} \mathbb{E}(X_i)}{\sum_{i=1}^{n} v_i} = \frac{\sum_{i=1}^{n} \theta v_i}{\sum_{i=1}^{n} v_i} = \theta$$

and that

$$\mathbb{E}(\hat{\theta}_{alt}) = \frac{1}{n} \sum_{i=1}^{n} \frac{\mathbb{E}(X_i)}{v_i} = \frac{1}{n} \sum_{i=1}^{n} \frac{\theta v_i}{v_i} = \theta$$

as well, so both are unbiased estimators. Next, we can compute

$$\mathbb{V}\mathrm{ar}(\hat{\theta}_{mle}) = \frac{\sum_{i=1}^{n} \mathbb{V}\mathrm{ar}(X_i)}{\left[\sum_{i=1}^{n} v_i\right]^2} = \frac{\sum_{i=1}^{n} \theta v_i}{\left[\sum_{i=1}^{n} v_i\right]^2} = \frac{\theta}{\sum_{i=1}^{n} v_i}$$

and

$$\mathbb{V}\mathrm{ar}(\hat{\theta}_{alt}) = \frac{1}{n^2} \sum_{i=1}^{n} \frac{\mathbb{V}\mathrm{ar}(X_i)}{v_i^2} = \frac{1}{n^2} \sum_{i=1}^{n} \frac{\theta v_i}{v_i^2} = \frac{\theta}{n^2} \sum_{i=1}^{n} \frac{1}{v_i}.$$

Since both are unbiased estimators, their respective mean squared errors depend solely on their variances, so which one has a smaller variance, or is more efficient? This can be determined by applying the Cauchy–Schwarz inequality (see Mathematical Insert 6 at the beginning of this section):

$$\left(\sum_{i=1}^{n} v_i\right)\left(\sum_{i=1}^{n} \frac{1}{v_i}\right) \geq \left(\sum_{i=1}^{n} 1\right)^2 = n^2 \quad \Rightarrow \quad \frac{1}{n^2}\left(\sum_{i=1}^{n} \frac{1}{v_i}\right) \geq \frac{1}{\left(\sum_{i=1}^{n} v_i\right)},$$

which means $\mathbb{V}\mathrm{ar}(\widehat{\theta}_{alt}) \geq \mathbb{V}\mathrm{ar}(\widehat{\theta}_{mle})$. [*Think: Intuitively, why is it "obvious" that $\widehat{\theta}_{alt}$ can be a rather unstable way to estimate the parameter?*] That is, the original, overall MLE is superior to the average of n "individual MLEs". In fact, one can also show in this case that $\widehat{\theta}_{mle}$ is the minimum-variance estimator among all linear and unbiased estimators of θ (Exercise 5.6). □

Exercise 5.6. Continue with Example 5.4. Consider the class of all linear and unbiased estimators of θ:

$$\mathcal{V}_\theta = \left\{\widehat{\theta}_a \equiv \sum_{i=1}^{n} a_i X_i \quad \text{s.t.} \quad \mathbb{E}(\widehat{\theta}_a) = \theta\right\}.$$

Show that $\widehat{\theta}_{mle}$ is the minimum-variance estimator in this class. [*Hint: Apply the Cauchy–Schwarz inequality. Alternatively, use the method of Lagrange multipliers (Mathematical Insert 9 in Appendix 5.A) to find explicitly the minimum-variance estimator in $\mathcal{V}_\theta$, and show that it coincides with $\widehat{\theta}_{mle}$.*] □

Exercise 5.7. Let $X_1, \ldots, X_n \overset{iid}{\sim} N(\mu, \sigma^2)$. Show that

$$\mathbb{E}\left[\sum_{i=1}^{n}(X_i - \bar{X})^2\right] = (n-1)\sigma^2$$

and hence conclude that $\widehat{\sigma}_{mle}^2$ is biased. [*Hint: First, show that the sum inside the square brackets is equal to $\sum_{i=1}^{n} X_i^2 - n\bar{X}^2$. Then, find $\mathbb{E}(X_i^2) = \mathbb{V}\mathrm{ar}(X_i) + [\mathbb{E}(X_i)]^2$ and, likewise, $\mathbb{E}(\bar{X}^2)$.*] □

Exercise 5.8. Let $X_1, \ldots, X_n \overset{iid}{\sim} \text{Uniform}(0, \theta)$.

(a) Find $\widehat{\theta}_{mle}$, the maximum likelihood estimator of θ. [*Hint: Given $X_1, \ldots, X_n$, there is an implicit constraint on the parameter θ. What is it?*]

(b) Find an alternative estimator, $\widehat{\theta}_{alt}$, based on the substitution principle (see Remark 5.1; section 5.1). [*Hint: What is the expectation of the Uniform$(0, \theta)$ distribution? See Exercise 2.7 in section 2.3.3.*]

(c) Compare $\widehat{\theta}_{mle}$ and $\widehat{\theta}_{alt}$ by following the steps of Exercise 4.1. [*Hint: Try repeating this with data sets of a few different sizes, for example, $n = 10, 100, and 1,000$.*]

(d) Compare $\hat{\theta}_{mle}$ and $\hat{\theta}_{alt}$ in terms of their respective biases, variances, and mean squared errors.

[*Think: This exercise shows that insisting on unbiasedness is not necessarily always the best.*]   □

5.3 Some advanced techniques

Reality is often complicated. In this section, we will continue with Example 5.2 to illustrate how a practical nuance can complicate the probability model that we use the describe the data-generating process and hence "force" us to use more advanced tools for parameter estimation.

Example 5.5. Recall Example 5.2 (section 5.1.2). Now, suppose that the X_is will not be directly observable. Instead, we will only be able to observe whether $X_i > 0$ or $X_i = 0$.

For example, drivers may not want to reveal to an insurance company that they've had a number of accidents in the past for this will surely increase their premium, but, to mitigate the risk of the insurance industry, government regulations may require that they have to reveal *something*. So a compromise is reached to balance privacy and transparency—drivers must reveal whether they've had any accident at all (i.e. whether $X_i > 0$ or $X_i = 0$), although they need not reveal exactly how many they've had.

Perhaps somewhat surprisingly, we can still estimate the parameter θ in this case, despite the added complication! The key insight here is that we can define another (fully observable) random variable,

$$Y_i = \begin{cases} 1, & X_i > 0, \\ 0, & X_i = 0. \end{cases}$$

Then, $Y_i \sim \text{Binomial}(1, p_i)$, where

$$p_i = \mathbb{P}(Y_i = 1) = \mathbb{P}(X_i > 0) = 1 - \mathbb{P}(X_i = 0)$$

$$= 1 - \frac{e^{-\theta v_i}(\theta v_i)^0}{0!} = 1 - e^{-\theta v_i}.$$

So, while we cannot write down a likelihood function based on the joint distribution of $X_1, \ldots, X_n$ because we won't have the exact value of $x_1, \ldots, x_n$, we can nonetheless write down a likelihood function based on the joint distribution of $Y_1, \ldots, Y_n$:

$$L(\theta) = \prod_{i=1}^{n}(1 - e^{-\theta v_i})^{y_i}(e^{-\theta v_i})^{1-y_i}.$$

This all sounds very exciting indeed. Unfortunately, there is no free lunch. Trying to maximize the corresponding log-likelihood function,

$$\ell(\theta) = \sum_{i=1}^{n} y_i \log(1 - e^{-\theta v_i}) + (1 - y_i)(-\theta v_i),$$

we discover that there is no analytic solution to the first-order condition,

$$\ell'(\theta) = \sum_{i=1}^{n} y_i \left[\frac{(-e^{-\theta v_i})(-v_i)}{1 - e^{-\theta v_i}} \right] - (1 - y_i)v_i = 0.$$

In cases like this, we will have to rely on numeric procedures to obtain a solution. A basic numeric procedure is Newton's algorithm (see Mathematical Insert 7 below), which iteratively updates the estimate from an initial guess $\widehat{\theta}_0$ by

$$\widehat{\theta}_{t+1} = \widehat{\theta}_t - \frac{\ell'(\widehat{\theta}_t)}{\ell''(\widehat{\theta}_t)}$$

for $t = 0, 1, 2, \ldots$ until convergence (e.g. $|\widehat{\theta}_{t+1} - \widehat{\theta}_t| < \varepsilon$), where ε is a small number, such as 1.0×10^{-10}, and t is an index for the iteration.[1]   ☐

Exercise 5.9. Continue Example 5.5 by working out the second derivative of the log-likelihood function, $\ell''(\theta)$, implementing the Newton's algorithm, and applying it to the data set given in Table 5.1.   ☐

Exercise 5.10. Suppose $X_i \sim \text{Binomial}(1, p_i)$, with

$$p_i = \sigma(v_i) \equiv \frac{\exp(\alpha + \beta v_i)}{1 + \exp(\alpha + \beta v_i)}, \tag{5.13}$$

independently for $i = 1, 2, \ldots, n$. Find the MLEs of both parameters α and β by implementing the Newton–Raphson algorithm (Mathematical Insert 8 below) and applying it to the data set given in Table 5.1. [*Think: The function*

---

[1] Here, we are referring to Newton's algorithm as a "basic" numeric procedure because its multivariate extension (see Mathematical Insert 8 below) has many variations, such as Gauss-Newton, quasi-Newton and so on. Numeric optimization is a very rich discipline on its own, and it is becoming ever more important for statistics and data science as we try to model more and more complicated data-generating processes.

Table 5.1 Number of individuals (out of 40) who had zero ($X_i = 0$) or at least one ($X_i > 0$) incident, along with their level of activity (v_i)

v_i	One + ($X_i > 0$)	Zero ($X_i = 0$)
8	10	0
4	8	2
2	7	3
1	3	7

Source: authors.

$\sigma(\cdot)$, given in Equation (5.13), is called the "sigmoid function" or the "sigmoid transform"; it is used widely in neural network models as a neuron activation function. Why is it important to use such a transform here, as opposed to, for example, a simple linear transform, $p_i = \alpha + \beta v_i$?] □

5.3.1 EM algorithm

By now, it is clear that the difficulty in Example 5.5 was caused by some of the X_is not being fully observable. This is sometimes known in statistics as a *missing data problem*.

Let us partition the data set D into an observable part D_{obs} and an unobservable part D_{mis}. When applying the method of maximum likelihood, we aim to solve:

$$\max_{\theta} \quad \ell(\theta; D_{obs}, D_{mis})$$

but we are "stuck" because solving the optimization problem above requires that we have access to D_{mis} but we don't.

Mathematical Insert 7

Newton's algorithm. To solve an equation, $f(x) = 0$, Newton's algorithm starts with an initial guess, say x_0, and iteratively updates the guess by:

$$x_{t+1} = x_t - \frac{f(x_t)}{f'(x_t)} \tag{5.14}$$

until convergence. The updating equation above is most easily seen graphically (see Figure 5.2). To optimize a function, $f(x)$, we'd like to solve

continued

continued

---

for the first-order condition, $f'(x) = 0$, so the updating equation above becomes:

$$x_{t+1} = x_t - \frac{f'(x_t)}{f''(x_t)}. \tag{5.15}$$

A "caveat" is that the iterative algorithm can only converge to a local solution when the initial guess x_0 is "reasonably close"; the algorithm may diverge if the initial guess is "bad". The multivariate extension is known as the Newton–Raphson algorithm (see Mathematical Insert 8 below).

Remark 5.6. One of the earliest known algorithms for finding $\sqrt{a}$ works as follows: repeatedly divide the current guess (x_t) into the number a and average the result with it to obtain a better guess (x_{t+1}), that is,

$$x_{t+1} = \frac{1}{2}\left(x_t + \frac{a}{x_t}\right),$$

until the guess stabilizes. It can be traced back to the early Babylonians and Greeks. Interestingly, it amounts to solving the equation $x^2 - a = 0$ by Newton's method! (Verify it.)

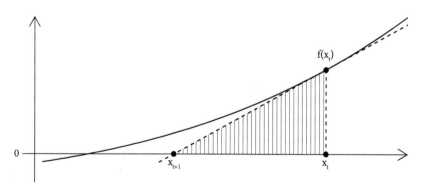

Figure 5.2   Explaining Newton's algorithm for solving $f(x) = 0$.

Note: The shaded triangle encodes the relation, $f'(x_t) = f(x_t)/(x_t - x_{t+1})$, which, after rearranging, gives the updating equation (5.14).

Source: authors.

Mathematical Insert 8

Newton–Raphson algorithm. To optimize a multivariate function, $f(x)$, where $x \in \mathbb{R}^d$, the corresponding updating equation that generalizes (5.15) is:

$$x_{t+1} = x_t - H_t^{-1} g_t,$$

where

$$H_t = \begin{bmatrix} \frac{\partial^2}{\partial x_1^2} f(x) & \frac{\partial^2}{\partial x_1 \partial x_2} f(x) & \cdots & \frac{\partial^2}{\partial x_1 \partial x_d} f(x) \\ \frac{\partial^2}{\partial x_2 \partial x_1} f(x) & \frac{\partial^2}{\partial x_2^2} f(x) & \cdots & \frac{\partial^2}{\partial x_2 \partial x_d} f(x) \\ \vdots & \vdots & \ddots & \vdots \\ \frac{\partial^2}{\partial x_d \partial x_1} f(x) & \frac{\partial^2}{\partial x_d \partial x_2} f(x) & \cdots & \frac{\partial^2}{\partial x_d^2} f(x) \end{bmatrix}_{x = x_t}$$

is the Hessian matrix and

$$g_t = \begin{bmatrix} \frac{\partial}{\partial x_1} f(x) \\ \vdots \\ \frac{\partial}{\partial x_d} f(x) \end{bmatrix}_{x = x_t}$$

the gradient vector of f, both evaluated at x_t. The Hessian matrix is expensive to compute for large d (i.e. high-dimensional problems), and there are many variations of the Newton–Raphson algorithm that try to deal with this particular difficulty in different ways.

A naturally logical reaction to this dilemma is to ask whether we can replace the missing component D_{mis} with something else, for example, perhaps an educated guess? This strategy actually works—*provided that* we carefully define what we mean by an "educated guess". The resulting algorithm is called the *EM algorithm*.

Like Newton's algorithm, the EM algorithm also works by iteratively updating an initial guess, $\hat{\theta}_0$, except that its updating principles are somewhat different. In particular, each update consists of two steps: an E-step, where "E" stands for "expectation"; and an M-step, where "M" stands for "maximization".

In the E-step, we compute

$$Q(\theta; D_{obs}, \hat{\theta}_{t-1}) \equiv \mathbb{E}_{D_{mis}|D_{obs}; \hat{\theta}_{t-1}} [\ell(\theta; D_{obs}, D_{mis})]. \qquad (5.16)$$

This is what we mean precisely by taking an educated guess. The log-likelihood function $\ell(\theta; D_{obs}, D_{mis})$ contains missing (i.e. unknown) components, which prevents us from being able to maximize it directly over θ. So, we start by taking an educated guess of what $\ell(\theta; D_{obs}, D_{mis})$ is. To do so, we ask for its expectation with respect to the conditional distribution of D_{mis} given D_{obs}; that is, our educated guess of whatever is *not* observable is informed by whatever *is* observable. Moreover, when taking this (conditional) expectation, we will need to rely on the current parameter estimate, $\widehat{\theta}_{t-1}$, as well. That's what the notation "$\mathbb{E}_{D_{mis}|D_{obs};\widehat{\theta}_{t-1}}(\cdot)$" means.

In the M-step, we compute

$$\widehat{\theta}_t \;=\; \arg\max_{\theta} \;\; Q(\theta; D_{obs}, \widehat{\theta}_{t-1}). \tag{5.17}$$

As we have just explained, the function $Q(\theta; D_{obs}, \widehat{\theta}_{t-1})$ is our educated guess of what the log-likelihood function $\ell(\theta; D_{obs}, D_{mis})$ is, based on whatever *is* observable (namely, D_{obs}) and our current guess of the parameter (namely, $\widehat{\theta}_{t-1}$). So, now we simply maximize it over θ to update the parameter estimate. This step is the same as what we would "usually" do when applying the maximum likelihood method if there were no missing components.

It is absolutely necessary to look at a concrete example in order to flush out what all this means.

Example 5.6. Recall Examples 5.2 and 5.5. The original log-likelihood was given in Equation (5.7), but we will copy and paste it below for convenience, while dropping all terms not involving θ:

$$\ell(\theta) = \sum_{i=1}^{n} -\theta v_i + X_i \log(\theta v_i).$$

We have also used X_i rather than x_i above because we no longer have observed values of $x_1, \ldots, x_n$ anymore. With some of these X_is not being fully observable, we are unable to directly maximize the function above over θ, so let's take an educated guess of this entire quantity. Following the E-step, we compute

$$Q(\theta; D_{obs}, \widehat{\theta}_{t-1}) \;=\; \mathbb{E}_{D_{mis}|D_{obs};\widehat{\theta}_{t-1}}\left[\sum_{i=1}^{n} -\theta v_i + X_i \log(\theta v_i)\right]$$

$$=\; \sum_{i=1}^{n} -\theta v_i + [\mathbb{E}_{D_{mis}|D_{obs};\widehat{\theta}_{t-1}}(X_i)] \log(\theta v_i). \tag{5.18}$$

Therefore, the E-step here simply amounts to replacing each X_i in the log-likelihood function with $\mathbb{E}_{D_{mis}|D_{obs};\widehat{\theta}_{t-1}}(X_i)$, our educated guess.

If $X_i = 0$ is observable, then clearly we need not replace it with anything; we only need an active replacement (guess) when $X_i > 0$ and the exact value of X_i is not observable. That is,

$$
\mathbb{E}_{D_{mis}|D_{obs};\widehat{\theta}_{t-1}}(X_i) =
\begin{cases}
\mathbb{E}(\underset{\substack{\uparrow \\ D_{mis}}}{X_i} \mid \underbrace{X_i > 0}_{D_{obs}}; \widehat{\theta}_{t-1}), & X_i > 0, \\[2em]
0, & X_i = 0.
\end{cases}
$$

The fact that $X_i > 0$ is, in itself, an observable piece of information, which should certainly be used when we try to make our educated guess of what X_i should be equal to on average. We will leave it as an exercise (Exercise 5.11) for you to verify that

$$
\mathbb{E}(X_i|X_i > 0; \widehat{\theta}_{t-1}) = \frac{\widehat{\theta}_{t-1}v_i}{1 - e^{-\widehat{\theta}_{t-1}v_i}}.
$$

In the M-step, we maximize (5.18) over θ. Notice that, here, the function $Q(\theta; D_{obs}, \widehat{\theta}_{t-1})$ looks almost exactly the same as the original log-liklihood function $\ell(\theta)$, except that each X_i has been replaced by $\mathbb{E}_{D_{mis}|D_{obs};\widehat{\theta}_{t-1}}(X_i)$. Thus, maximizing it gives

$$
\widehat{\theta}_t = \frac{\sum_{i=1}^{n}\mathbb{E}_{D_{mis}|D_{obs};\widehat{\theta}_{t-1}}(X_i)}{\sum_{i=1}^{n}v_i} = \frac{\sum_{X_i>0}(\widehat{\theta}_{t-1}v_i)/(1 - e^{-\widehat{\theta}_{t-1}v_i})}{\sum_{i=1}^{n}v_i},
$$

an iterative updating equation from $\widehat{\theta}_{t-1}$ to $\widehat{\theta}_t$.   □

Exercise 5.11. Let $X \sim \text{Poisson}(\lambda)$. Show that

$$
\mathbb{E}(X|X > 0) = \frac{\lambda}{1 - e^{-\lambda}}.
$$

[*Hint: First, find* $\mathbb{P}(X = x|X > 0)$ *for* $x \in \{1, 2, 3, \dots\}$.]   □

Exercise 5.12. Continue Examples 5.6 by implementing the EM algorithm and applying it to the data set given in Table 5.1. Does the EM algorithm produce the same numeric estimate as Newton's algorithm (Exercise 5.9)? Which algorithm converges faster?   □

5.3.2 Latent variables

Even though the EM algorithm has been presented above as a special strategy to deal with missing data problems, its applicability is actually more general than that. Sometimes, we could describe the data-generating process more easily if we introduced extra *latent variables*, that is, imaginary or non-existent (and hence unobservable) quantities that nonetheless help us better describe the data-generating process.

For example, when situations are complex enough, it may be difficult for us to find a single probability model to describe the data-generating process producing the random variables $X_1, \ldots, X_n$. Under such circumstances, we can postulate that the data-generating process may consist of two stages: first, a (hidden) label $Z_i \in \{1, \ldots, K\}$ is generated (from a multinomial distribution); then, conditional on the label $Z_i = k$, a random variable X_i is generated from a relatively simple model, say $f(x; \theta_k)$.

So, effectively, a total of K different probability models are at work (rather than just one), and each X_i can be generated by any of these K models with different probabilities. This is certainly a much richer model that is capable of describing phenomena that any one of these models cannot do on its own; it is called a *finite mixture model*. These (hidden) labels $Z_1, \ldots, Z_n$ are crucial for allowing us to describe such a two-stage data-generating process with clarity, but they themselves will remain unobservable because their existence is based purely upon our own imagination. The EM algorithm allows us to estimate such a model, by treating $\{X_1, \ldots, X_n\}$ as D_{obs} and $\{Z_1, \ldots, Z_n\}$ as D_{mis}. Interested readers may refer to Appendix 5.A for some more details.

Remark 5.7. The stochastic block model (Example 1.1 at the beginning of Chapter 1) and its continuous-time extension (Example 1.2) are, in fact, latent-variable models of this kind, but they are more complicated to estimate than the finite-mixture models mentioned here. For the stochastic block model [Equation (1.1)], in order to generate the random variable X_{ij}, one must first generate two hidden labels, Z_i and Z_j, rather than just one. Exactly why this makes the model harder to estimate can be somewhat subtle for students to appreciate; we will not go into it here. □

Appendix 5.A Finite mixture models

According to the two-stage data-generating process (section 5.3.2), the joint distribution for each pair, (X_i, Z_i), is

$$f(x_i, z_i) = f(x_i|z_i)\, f(z_i) = \prod_{k=1}^{K} [\underbrace{f(x_i|z_i = k)}_{f(x_i;\theta_k)}\, \underbrace{\mathbb{P}(Z_i = k)}_{\pi_k}]^{I(z_i = k)}$$

where $\overset{\uparrow}{\mathbb{P}(Z_i = z_i)}$

$$= \prod_{k=1}^{K} [f(x_i; \theta_k)\pi_k]^{I(z_i = k)},$$

where $I(\cdot)$ is a binary indicator function such that

$$I(A) = \begin{cases} 1, & \text{if } A \text{ is true;} \\ 0, & \text{if } A \text{ is false.} \end{cases}$$

Recall that there are K different probability models at work. What the indicator function $I(z_i = k)$ does is to pick out the correct probability model for (X_i, Z_i) based on the label Z_i. Here, it is important to take note of an implicit constraint on the parameters $\pi_1, \ldots, \pi_K$, namely,

$$\sum_{k=1}^{K} \pi_k = 1. \tag{5.19}$$

Taking a product over all pairs $(x_1, z_1), \ldots, (x_n, z_n)$ gives the overall likelihood function:

$$L(\theta_1, \ldots, \theta_K, \pi_1, \ldots, \pi_K) = \prod_{i=1}^{n} f(x_i, z_i) = \prod_{i=1}^{n} \prod_{k=1}^{K} [f(x_i; \theta_k)\pi_k]^{I(z_i = k)},$$

except that, of course, we don't actually have observed values of $z_1, \ldots, z_n$. To simplify the notation, we will use

$$\Theta = \{\theta_1, \ldots, \theta_K, \pi_1, \ldots, \pi_K\}$$

to denote the set of all parameters;

$$X = \{X_1, \ldots, X_n\} \text{ and interchangeably } \{x_1, \ldots, x_n\}$$

to denote the set of all observable random variables or their actual observed values; and

$$\mathbf{Z} = \{Z_1, \ldots, Z_n\}$$

to denote the set of all unobservable random variables, which we postulated when describing the data-generating process. Then, the log-likelihood function is

$$\ell(\mathbf{\Theta}; X, Z) = \sum_{i=1}^{n} \sum_{k=1}^{K} [I(Z_i = k)] \log[f(x_i; \theta_k)\pi_k],$$

and the previously general notation "$\mathbb{E}_{D_{mis}|D_{obs};\hat{\theta}_{t-1}}(\cdot)$" now means, more specifically, $\mathbb{E}_{Z|X;\hat{\Theta}_{t-1}}(\cdot)$. Thus,

$$Q(\mathbf{\Theta}; X, \hat{\Theta}_{t-1}) = \mathbb{E}_{Z|X;\hat{\Theta}_{t-1}}[\ell(\mathbf{\Theta}; X, Z)]$$

$$= \sum_{i=1}^{n} \sum_{k=1}^{K} \underbrace{\mathbb{E}_{Z|X;\hat{\Theta}_{t-1}}[I(Z_i = k)]}_{w_{ik}} \log[f(x_i; \theta_k)\pi_k]$$

$$= \sum_{i=1}^{n} \sum_{k=1}^{K} w_{ik} \log[f(x_i; \theta_k)] + w_{ik} \log(\pi_k).$$

We will leave it as an exercise (Exercise 5.13) for you to compute w_{ik} and hence complete the E-step.

Given w_{ik}, the M-step is relatively straightforward. Updating each θ_k simply requires us to solve a weighted MLE problem,

$$\max_{\theta_k} \sum_{i=1}^{n} w_{ik} \log[f(x_i; \theta_k)].$$

Due to the constraint (5.19), however, updating the parameters $\pi_1, \ldots, \pi_K$ requires us to solve a *constrained* MLE problem,

$$\max_{\pi_1, \ldots, \pi_K} \sum_{k=1}^{K} w_{\cdot k} \log(\pi_k) \quad \text{s.t.} \quad \sum_{k=1}^{K} \pi_k = 1,$$

where $w_{\cdot k} \equiv w_{1k} + w_{2k} + \ldots + w_{nk}$. This is an optimization problem with an *equality constraint*, which can be solved with the method of Lagrange multipliers (see Mathematical Insert 9 below). The necessary conditions are:

$$\frac{d}{d\pi_k}\left[\sum_{k=1}^{K} w_{\cdot k}\log \pi_k + \lambda\left(\sum_{k=1}^{K} \pi_k - 1\right)\right] = 0 \quad \text{for all} \quad k = 1, 2, \ldots, K;$$

$$\frac{d}{d\lambda}\left[\sum_{k=1}^{K} w_{\cdot k}\log \pi_k + \lambda\left(\sum_{k=1}^{K} \pi_k - 1\right)\right] = 0.$$

Solving the first equation above gives

$$\frac{w_{\cdot k}}{\pi_k} + \lambda = 0 \quad \Rightarrow \quad \pi_k = \frac{w_{\cdot k}}{(-\lambda)} \quad \text{for all} \quad k = 1, 2, \ldots, K.$$

Plugging this into the second one gives

$$\sum_{k=1}^{K} \pi_k = 1 \quad \Rightarrow \quad \sum_{k=1}^{K} \frac{w_{\cdot k}}{(-\lambda)} = 1 \quad \Rightarrow \quad (-\lambda) = \sum_{k=1}^{K} w_{\cdot k} \equiv w_{\cdot\cdot}.$$

Thus, the MLE is $\hat{\pi}_k = w_{\cdot k}/w_{\cdot\cdot}$ for every k.

Exercise 5.13. Complete the EM algorithm for fitting finite-mixture models by showing that

$$w_{ik} = \mathbb{P}(Z_i = k | X_i = x_i; \hat{\Theta}) = \frac{\hat{\pi}_k f(x_i; \hat{\theta}_k)}{\hat{\pi}_1 f(x_i; \hat{\theta}_1) + \ldots + \hat{\pi}_K f(x_i; \hat{\theta}_K)},$$

where we have suppressed the subscript "$t-1$" by simply writing "$\hat{\Theta}$" for "$\hat{\Theta}_{t-1}$". [*Hint: Apply Bayes law. Note the denominator above also suggests the marginal distribution of each X_i is*

$$f(x_i) = \sum_{\text{all } z_i} f(x_i, z_i) = \sum_{\text{all } z_i} f(z_i)f(x_i|z_i) = \sum_{k=1}^{K} \pi_k f(x_i; \theta_k).$$

Why?]   □

Mathematical Insert 9

Method of Lagrange multipliers. Let $x = (x_1, x_2, \ldots, x_d)^\top \in \mathbb{R}^d$. To solve an optimization problem with *equality constraints*,

$$\max_{x} \quad f(x)$$

$$\text{s.t.} \quad g_1(x) = 0,$$

$$g_2(x) = 0,$$

$$\vdots$$

$$g_m(x) = 0,$$

the *necessary conditions* for the solution are:

$$\frac{d}{dx_j}\left[f(x) + \sum_{k=1}^{m} \lambda_k g_k(x) \right] = 0 \quad \text{for each} \quad j = 1, 2, \ldots, d;$$

$$\frac{d}{d\lambda_k}\left[f(x) + \sum_{k=1}^{m} \lambda_k g_k(x) \right] = 0 \quad \text{for each} \quad k = 1, 2, \ldots, m.$$

The auxiliary variables $\lambda_1, \lambda_2, \ldots, \lambda_m$, introduced inside the square brackets above to facilitate the solution of this problem, are referred to as Lagrange multipliers.

Remark 5.8. Interestingly, although the Lagrange multiplier method dates back to the 1700s, necessary conditions for solving optimization problems with *inequality constraints* did not become available until well into the twentieth century. Today, they are known as KKT conditions, named after their discoverers: William Karush, Harold Kuhn, and Albert Tucker.

Exercise 5.14. Based on data collected from *The London Times*, the number of days during the period of 1910–1912 with x number of deaths to women ≥ 80 years of age, for $x = 0, 1, \ldots, 10+$, are tabulated in Table 5.2. Let X be a random variable for the number of deaths per day to women in this age group. Using these data, implement an EM algorithm to fit a mixed Poisson model,

Table 5.2 Data from *The London Times* for the period 1910–1912

x	0	1	2	3	4	5	6	7	8	9	10+
#{days}	162	267	271	185	111	61	27	8	3	1	0

Source: authors.

$$\mathbb{P}(X = x) = \pi \left[\frac{e^{-\lambda_1} \lambda_1^x}{x!} \right] + (1 - \pi) \left[\frac{e^{-\lambda_2} \lambda_2^x}{x!} \right],$$

for the distribution of X. [*Think: How would you interpret the parameters λ_1, λ_2 and π?*][2]   □

Remark 5.9. Earlier in Chapter 3 (see Remark 3.1 in section 3.1), we briefly mentioned that perhaps the two most ubiquitous probability models encountered in practice are the multivariate normal distribution, for real-valued data, and the multinomial distribution, for categorical data. Not surprisingly, finite mixtures of these distributions also are widely used—that is, finite mixtures in which each individual component $f(x_i; \theta_k)$ is Normal(μ_k, Σ_k) or Multinomial($m_i; p_k$). If x is d-dimensional, then μ_k is a vector in $\mathbb{R}^d$; Σ_k is a $d \times d$ positive-definite matrix; and $p_k = (p_{1k}, p_{2k}, \ldots, p_{dk})^\top$ with $p_{jk} \geq 0 \; \forall \; j$ and $p_{1k} + p_{2k} + \ldots + p_{dk} = 1$.

Mixtures of multivariate normals are frequently used to *cluster* multidimensional data, while mixtures of multinomials form the baseline of various *topic models* for text data. When text data are modeled by the multinomial distributions (or their mixtures), a dictionary is first chosen which contains d most commonly used words, and each piece of text is then represented by a vector $x_i = (x_{i1}, x_{i2}, \ldots, x_{id})^\top$, in which the component x_{ij} simply counts the number of times a particular word (here, indexed by j) occurs in the text (here, indexed by i). In this type of model, the length of each text, $m_i \equiv x_{i1} + x_{i2} + \ldots + x_{id}$, is usually treated as a fixed quantity and *not* modeled by a probabilistic generating mechanism.

One can easily appreciate why it is attractive to use mixture models for text data. An article about Renaissance art will undoubtedly use a very different set of words more frequently than an article about molecular biology. Thus, the two-stage data-generating process is much more realistic—although still far from being *truly* realistic—for a rich data set of different texts. First, a hidden topic is generated; given the topic (here, indexed by k), the words

---

[2] This exercise has been adapted from the text by Kenneth Lange [8].

are then generated from a multinomial distribution with a topic-specific parameter, $\boldsymbol{p}_k$.

However, fitting these mixture models in high dimensions is already quite a tricky problem. For example, the temptation to fit the most flexible normal-mixture model is technically not advisable [9]. Often, some restrictions on the matrices $\boldsymbol{\Sigma}_1, \boldsymbol{\Sigma}_2, \ldots, \boldsymbol{\Sigma}_K$ are necessary, the simplest kind being $\boldsymbol{\Sigma}_k = \sigma^2 \boldsymbol{I}$ for all k. (See also Exercise 5.15 below.) For multinomial mixtures, the parameters

$$\{p_{jk} : j = 1, 2, \ldots, d; k = 1, 2, \ldots, K\}$$

are usually hard to estimate accurately since most words simply do not appear very often, especially if the size of the data set is not all that large (e.g. $5,000$ articles) compared with the size of the dictionary (e.g. $d = 20,000$ words). That's why many variations have been proposed [10], and the "vanilla" mixture model we described here is seldom sufficient in reality.

□

Exercise 5.15. How do you implement the EM algorithm for fitting a finite mixture of multivariate normal distributions, in which each component $f(\boldsymbol{x}; \boldsymbol{\theta}_k)$ is taken to be $N(\boldsymbol{\mu}_k, \sigma^2 \boldsymbol{I})$ with $\sigma \to 0$? [*Note: The resulting algorithm becomes the well-known "K means" algorithm for clustering.*]   □

6

Bayesian Approach

This chapter focuses on the Bayesian approach to "saying something about" the model parameters—specifically, how to find their posterior distribution.

6.1 Basics

It is best to start with a concrete example.

Example 6.1. Let's consider the simple case of $X_1, \ldots, X_n \overset{iid}{\sim}$ Poisson(λ). Recall from Example 5.1 in section 5.1.1 earlier that the MLE is simply $\widehat{\lambda}_{mle} = \bar{X}$. Now, to proceed as a Bayesian, we must first specify a prior distribution, $\pi(\lambda)$, for λ. We will choose

$$\lambda \sim \text{Gamma}(\alpha, \beta)$$

and explain later why this is a favorite choice in this particular context. Having made our choice of the prior, we can now proceed to obtain the posterior,

$$\pi(\lambda|x_1, \ldots, x_n) = \frac{f(x_1, \ldots, x_n|\lambda)\pi(\lambda)}{\displaystyle\int f(x_1, \ldots, x_n|\lambda)\pi(\lambda)\, d\lambda}.$$

To do so, the most difficult piece of calculation is to compute the marginal distribution of $(X_1, \ldots, X_n)$ in the denominator—a compound distribution (see section 3.2.2). Here, we have

$$
\begin{aligned}
m(x_1, \ldots, x_n) &= \int f(x_1, \ldots, x_n|\lambda)\pi(\lambda)\, d\lambda \\
&= \int \left[\prod_{i=1}^{n} \frac{e^{-\lambda}\lambda^{x_i}}{x_i!} \right] \times \left[\frac{\beta^\alpha}{\Gamma(\alpha)} \lambda^{\alpha-1} e^{-\beta\lambda} \right] d\lambda.
\end{aligned}
$$

Usually, this integral would be hard, if not impossible, to evaluate analytically. It is at this stage that we begin to see more clearly why our earlier choice of

Essential Statistics for Data Science. Mu Zhu, Oxford University Press. © Mu Zhu (2023).
DOI: 10.1093/oso/9780192867735.003.0006

Gamma(α, β) as the prior for λ is extremely attractive. In terms of λ, the Poisson mass function and the Gamma density function have the same form—a polynomial term in λ and an exponential term. This allows us to use our favorite technique for integration—namely, we can make the integrand into another Gamma density function and exploit the fact that all density functions must integrate to one. In particular, continuing from where we left off, we obtain

$$
m(x_1, \ldots, x_n) = \frac{\beta^\alpha}{\Gamma(\alpha) \prod_{i=1}^{n} x_i!} \times \frac{\Gamma(n\bar{x} + \alpha)}{(n + \beta)^{n\bar{x}+\alpha}} \times
$$

$$
\int \underbrace{\frac{(n + \beta)^{n\bar{x}+\alpha}}{\Gamma(n\bar{x} + \alpha)} \lambda^{n\bar{x}+\alpha-1} e^{-(n+\beta)\lambda}}_{\text{Gamma}(n\bar{x}+\alpha, n + \beta)} \, d\lambda
$$

$$
= \frac{\Gamma(n\bar{x} + \alpha)}{\Gamma(\alpha) \prod_{i=i}^{n} x_i!} \left[\frac{\beta}{n + \beta} \right]^\alpha \left[\frac{1}{n + \beta} \right]^{n\bar{x}}. \tag{6.1}
$$

Notice that this technique also immediately implies that

$$
\lambda | X_1, \ldots, X_n \sim \text{Gamma}(n\bar{X} + \alpha, n + \beta). \tag{6.2}
$$

(Why? If this does not yet seem obvious to you, it may be a good idea to review Example 3.4 again.) A prior distribution that "matches up" like this with the probability model is called a *conjugate prior*.[1]

If we want to produce a single estimate of the parameter (rather than an entire posterior distribution), we can—but are by no means obliged to—use the posterior expectation,

$$
\mathbb{E}(\lambda | X_1, \ldots, X_n) = \frac{n\bar{X} + \alpha}{n + \beta} = \frac{\bar{X} + \alpha/n}{1 + \beta/n}.
$$

If we compare this with the MLE, $\hat{\lambda}_{mle} = \bar{X}$, we see that, when the sample size n is relatively large, the Bayesian and frequentist conclusions will largely agree since, for fixed α, β the posterior expectation tends to $\bar{X}$ as $n \to \infty$. Thus, in

---

[1] Some other examples include: for Binomial(n, p), the conjugate prior for p is the Beta distribution; for N(μ, σ^2), the conjugate prior for μ is another Normal distribution; and so on.

this case, the choice of prior parameters—here, α and β—only has a negligible impact. □

Remark 6.1. Notice that Equation (6.1) does not factor. This means that, marginally, the random variables $X_1, X_2, \ldots, X_n$ are no longer independent; they are only *conditionally independent*, given λ. (See also Exercise 3.7 in section 3.2.2.) If α is an integer, then the marginal distribution of $(X_1, \ldots, X_n)$, which can be written as

$$m(x_1, \ldots, x_n) = \frac{(\sum_{i=1}^{n} x_i + \alpha - 1)!}{[(\alpha - 1)!](x_1!) \ldots (x_n!)} \left[\frac{\beta}{n+\beta}\right]^{\alpha} \left[\frac{1}{n+\beta}\right]^{x_1} \cdots \left[\frac{1}{n+\beta}\right]^{x_n},$$

is a so-called *negative multinomial distribution*. [*Think: Recall the multinomial distribution (Example 3.3 in section 3.1). Try to develop an intuition of why X_j and X_k can't be independent of each other, for any $1 \le j, k \le K$, if $(X_1, X_2, \ldots, X_K)$ jointly follow the Multinomial(n; $p_1, p_2, \ldots, p_K$) distribution. For the negative multinomial distribution, the intuition is similar in spirit but requires an understanding of the physical phenomenon that the negative multinomial distribution describes. However, this particular detail is largely irrelevant for our discussion here, so we choose to skip it.*] □

The prior distribution of θ,

$$\theta \sim \pi(\theta; \psi),$$

will inevitably also have some parameters, denoted here above by ψ; they are usually referred to as *hyperparameters*. (In Example 6.1, the hyperparameters are α, β of the Gamma distribution.) Clearly, the posterior distribution depends on ψ as well. When this fact needs to be emphasized, we will write

$$\pi(\theta|x_1, x_2, \ldots, x_n; \psi),$$

rather than just $\pi(\theta|x_1, x_2, \ldots, x_n)$. While we *may* prefer to simply choose a conjugate prior for θ whenever possible, that doesn't mean all the work is done; we still have to specify the hyperparameter ψ itself.

Thus, strictly speaking, the Bayesian approach does not really answer the question of what we can say about θ based on $\{X_1, X_2, \ldots, X_n\}$ alone (see Chapter 4); rather, it answers the question of what we can say about θ based on both $\{X_1, X_2, \ldots, X_n\}$ and ψ.

6.2 Empirical Bayes

For a class of simultaneous estimation problems, there is an interesting way to determine the hyperparameters empirically. The typical structure of these problems is as follows:

$$
\begin{aligned}
X_i|\theta_i &\sim f(x|\theta_i) \quad \text{independently for } i = 1, 2, \ldots, n, \\
\theta_i &\overset{iid}{\sim} \pi(\theta; \psi).
\end{aligned} \tag{6.3}
$$

That is, the random variables $X_1, X_2, \ldots, X_n$ are assumed to independently follow the same parametric family of distributions, but *each* X_i has its *own* unique parameter θ_i. In the Bayesian approach, these parameters are treated as independent random variables themselves; here, they are assumed to have the *same* prior distribution with hyperparameter ψ.

Then, it follows that, marginally (i.e. after integrating out θ_i), $X_1, X_2, \ldots, X_n$ are still independent and each follows the same marginal distribution,

$$
m(x; \psi) = \int f(x|\theta)\pi(\theta; \psi)d\theta.
$$

This means that, instead of specifying the hyperparameter ψ subjectively, we can actually estimate it from the sample, $\{X_1, X_2, \ldots, X_n\}$, for example, by maximizing the likelihood,

$$
L(\psi; x_1, x_2, \ldots, x_n) = \prod_{i=1}^{n} m(x_i; \psi).
$$

Let $\widehat{\psi}$ denote the corresponding estimate. We then plug it into the posterior to obtain $\pi(\theta_i|x_i; \widehat{\psi})$. This approach is known as the *empirical Bayes* approach; it is a kind of hybrid approach, combining both Bayesian and frequentist principles.

Example 6.2. Suppose

$$
\begin{aligned}
X_i|\lambda_i &\sim \text{Poisson}(\lambda_i) \quad \text{independently for } i = 1, 2, \ldots, n, \\
\lambda_i &\overset{iid}{\sim} \text{Gamma}(1, \theta).
\end{aligned} \tag{6.4}
$$

Here, we are deliberately using a slightly simpler Gamma prior (fixing the first parameter to be 1) so that there is only one hyperparameter (i.e. θ) to deal with, rather than two.

First, for each individual i, it is easy to see that this is merely a special case of what was considered in Example 6.1. Thus, by taking $n = 1$, $\alpha = 1$, and $\beta = \theta$, Equation (6.2) implies the posterior of $\lambda_i|X_i$ is simply

$$\lambda_i|X_i \quad \sim \quad \text{Gamma}(X_i + 1, 1 + \theta)$$

with posterior expectation $\mathbb{E}(\lambda_i|X_i) = (X_i + 1)/(1 + \theta)$, while Equation (6.1) says the marginal of X_i is simply

$$m(x_i) \quad = \quad \underbrace{\frac{\Gamma(x_i + 1)}{\Gamma(1)x_i!}}_{=1} \left[\frac{\theta}{1+\theta}\right]\left[\frac{1}{1+\theta}\right]^{x_i}$$

for all $i = 1, 2, \ldots, n$.

Instead of specifying an arbitrary value for θ, we can now estimate it from the marginal distributions of $X_1, \ldots, X_n$. The likelihood function is

$$L(\theta; x_1, \ldots, x_n) = \prod_{i=1}^{n} m(x_i) = \left[\frac{\theta}{1+\theta}\right]^{n}\left[\frac{1}{1+\theta}\right]^{\sum_{i=1}^{n} x_i}$$

and maximizing the log-likelihood

$$\ell(\theta) = n\log(\theta) - \left(n + \sum_{i=1}^{n} x_i\right)\log(1 + \theta)$$

gives

$$\ell'(\theta) = \frac{n}{\theta} - \frac{n + \sum_{i=1}^{n} x_i}{1 + \theta} = 0 \quad \Rightarrow \quad \widehat{\theta} = \frac{n}{\sum_{i=1}^{n} x_i} = \frac{1}{\bar{x}}.$$

Fixing the hyperparameter θ at $\widehat{\theta} = 1/\bar{x}$, the posterior distribution $\pi(\lambda_i|x_i; \widehat{\theta})$ is simply $\text{Gamma}(x_i + 1, 1 + \widehat{\theta})$ for every i, and the corresponding posterior expectation becomes

$$\mathbb{E}(\lambda_i|X_i = x; \widehat{\theta}) = \frac{x + 1}{1 + \widehat{\theta}} = (x + 1)\left[\frac{\bar{x}}{\bar{x} + 1}\right]. \tag{6.5}$$

At first glance, this is by all means a very strange-looking equation for it implies that our knowledge of λ_i shouldn't just depend on the value of X_i alone; it should depend on the values of all $X_1, \ldots, X_n$. Of course, the intuition that only X_i contains information about λ_i would be correct if the model specification contained only the statement that $X_1, \ldots, X_n$ are independent, each with its own parameter λ_i. But our model here specifies additionally that $\lambda_1, \ldots, \lambda_n$ share the same prior distribution with a common parameter θ. $\square$

In Equation (6.3), that all of $X_1, X_2, \ldots, X_n$ are used to estimate each individual parameter θ_i is a signature of the empirical Bayes methodology. Perhaps the most famous example in statistics is the James–Stein estimator (see Fun Box 2 below).

Exercise 6.1. Remarkably, the problem considered in Example 6.2 can be approached by assuming a non-parametric prior, $\pi(\lambda)$, for each λ_i as well. The marginal distribution of X_i now becomes

$$m(x) = \mathbb{P}(X_i = x) = \int \mathbb{P}(X_i = x|\lambda_i)\pi(\lambda_i)d\lambda_i = \int \left[\frac{e^{-\lambda_i}\lambda_i^x}{x!} \right] \pi(\lambda_i)d\lambda_i.$$

(a) Show that the posterior expectation is

$$\mathbb{E}(\lambda_i|X_i = x) = (x+1)\left[\frac{m(x+1)}{m(x)} \right]. \tag{6.6}$$

(b) In a data set of $n = 100$ individuals (Table 6.1), some have had no accident while various others have had one, two, or three accidents during the past five years. Suppose Amy is one of those who had one accident ($X_{Amy} = 1$) and Bob is one of those who had two ($X_{Bob} = 2$). Propose a "sensible" way to estimate $m(x)$, apply it to this data set, and compare the two different empirical Bayes estimates, (6.5) and (6.6), respectively for Amy and for Bob.

[*Think: For both Amy and Bob, respectively, compare also the two empirical Bayes estimates with their corresponding MLEs, $\widehat{\lambda}_{Amy}^{(mle)}$ and $\widehat{\lambda}_{Bob}^{(mle)}$. What are the reasons why these empirical Bayes estimates may be more attractive than the MLEs?*] □

Table 6.1 Number of individuals (out of 100) who have had zero, one, two, or three accidents during the past five years

Zero ($X_i = 0$)	One ($X_i = 1$)	Two ($X_i = 2$)	Three ($X_i = 3$)	Total (n)
70	20	8	2	100

Source: authors.

Fun Box 2

The James–Stein estimator. If $X_1, \ldots, X_n$ are independent random variables, each distributed as $X_i \sim N(\theta_i, \sigma^2)$, then $\widehat{\theta}_i^{(mle)} = X_i$, and it does *not* seem plausible that any other estimator is even possible. Indeed, statisticians were shocked in the late 1950s and early 1960s when better estimators were discovered for $n > 3$. Assuming that σ^2 is known (so it need not be estimated), one such estimator, the James–Stein estimator, is given by

$$\widehat{\theta}_i^{(js)} = \bar{X} + \left[1 - \frac{(n-3)\sigma^2}{\sum_{i=1}^{n}(X_i - \bar{X})^2}\right](X_i - \bar{X}). \tag{6.7}$$

In a famous application [11], historical batting averages of different baseball players $(X_1, \ldots, X_n)$ were used to estimate an individualized "ability parameter" for each player $(\theta_1, \ldots, \theta_n)$. When these estimated ability parameters were used to predict future performances, $\widehat{\theta}_i^{(js)}$ did a lot better than $\widehat{\theta}_i^{(mle)}$. To each individual player, it must be puzzling (and maybe even infuriating) why *other* players' past performances should have anything to do with the estimate of *his* ability! It took statisticians about 15 years to realize that (6.7) can be regarded as an empirical Bayes estimator. Assuming the prior distribution of each θ_i to be $N(\mu, \tau^2)$, one can show (Exercise 6.2) that

$$\mathbb{E}(\theta_i | X_i) = \mu + \left[1 - \frac{\sigma^2}{\sigma^2 + \tau^2}\right](X_i - \mu)$$

and that the marginal distribution of each X_i is $N(\mu, \sigma^2 + \tau^2)$. We can then estimate μ and $1/(\sigma^2 + \tau^2)$ from all of $X_1, \ldots, X_n$. Plugging in a set of unbiased estimators,

$$\mathbb{E}(\bar{X}) = \mu \quad \text{and} \quad \mathbb{E}\left[\frac{n-3}{\sum_{i=1}^{n}(X_i - \bar{X})^2}\right] \overset{\star}{=} \frac{1}{\sigma^2 + \tau^2}$$

gives the expression in Equation (6.7). (The step marked by "$\star$" above requires knowledge about the *Inverse Gamma distribution*, which is not covered by this book.)

Exercise 6.2. Consider the problem inside Fun Box 2 above. Show that the posterior distribution of $\theta_i | X_i$ is normal with expectation

$$\mathbb{E}(\theta_i|X_i) = \left[\frac{\tau^2}{\sigma^2 + \tau^2}\right] X_i + \left[\frac{\sigma^2}{\sigma^2 + \tau^2}\right] \mu = \mu + \left[1 - \frac{\sigma^2}{\sigma^2 + \tau^2}\right](X_i - \mu)$$

and that the marginal distribution of X_i is $N(\mu, \sigma^2 + \tau^2)$. □

6.3 Hierarchical Bayes

Here is a very natural thought: if the fundamental Bayesian idea is to treat the unknown parameter θ as a random variable, why not treat the unknown hyperparameter ψ of the prior distribution as a random variable as well instead of worrying about how to specify or estimate its value?

Having asked such a question, one immediately sees that, in principle, we can do this recursively forever. The random variable ψ will need a prior, which will have its own unknown parameter ψ', which we can also treat as a random variable, which will need another prior, which will have its unknown parameter ψ'', and so on. This approach is called the *hierarchical Bayes* approach. In practice, it is clear that we *must* stop at some point eventually, which means the "last" hyperparameter must *not* be treated as a random variable; instead, its value will have to be determined either subjectively or using some exogenous information.

Mathematical Insert 10

The Gibbs sampler. If we want, but don't know how, to sample from a multivariate distribution, say, $f(z_1, z_2, \ldots, z_d)$, we can instead sample sequentially from the conditional distributions of each coordinate given the rest,

$$f(z_j|z_1, \ldots, z_{j-1}, z_{j+1}, \ldots, z_d), \quad j = 1, 2, \ldots, d, 1, 2, \ldots, d, \ldots,$$

one coordinate at a time, until the distribution stabilizes (see the Remark below)—called "burn in" in this context.

Remark 6.2. Appendix 6.A explains in more detail that (i) mathematically, the notion of "stabilization" here means that the underlying Markov chain has reached its *stationary distribution* and that (ii) the Gibbs sampler is a special case of a more general Monte Carlo algorithm called the Metropolis–Hastings algorithm.

Example 6.3. Consider Example 6.2 again, except we will now assume that the hyperparameter θ is also a random variable, with its own prior distribution—sometimes referred to as a *hyperprior*. Just as we have chosen the prior of λ_i to be conjugate to the distribution of X_i, for convenience, we will continue to choose the hypeprior of θ to be conjugate to the distribution of λ_i. Hence, we have the following hierarchical model:

$$
\begin{aligned}
X_i|\lambda_i &\sim \text{Poisson}(\lambda_i) \quad \text{independently for } i = 1, 2, \ldots, n, \\
\lambda_i|\theta &\overset{iid}{\sim} \text{Gamma}(1, \theta), \\
\theta &\sim \text{Gamma}(\alpha, \beta).
\end{aligned}
$$

Unfortunately, with the hierarchical set-up, the full posterior distribution,

$$
\pi(\lambda_1, \ldots, \lambda_n, \theta|x_1, \ldots, x_n)
$$

is now much harder to obtain, even though conjugate priors are used at every level of the hierarchy.

To obtain the marginal (compound) distribution of $X_1, \ldots, X_n$, we must now integrate over all of $\lambda_1, \ldots, \lambda_n$ and θ:

$$
\begin{aligned}
&m(x_1, \ldots, x_n) \\
&= \int f(x_1, \ldots, x_n|\lambda_1, \ldots, \lambda_n) \times f(\lambda_1, \ldots, \lambda_n|\theta) \times \pi(\theta) \, d\theta d\lambda_1 \ldots d\lambda_n \\
&= \int \left[\prod_{i=1}^{n} \frac{e^{-\lambda_i}\lambda_i^{x_i}}{x_i!} \right] \left[\prod_{i=1}^{n} \theta e^{-\theta\lambda_i} \right] \left[\frac{\beta^\alpha}{\Gamma(\alpha)} \theta^{\alpha-1} e^{-\beta\theta} \right] d\theta d\lambda_1 \ldots d\lambda_n. \quad (6.8)
\end{aligned}
$$

Let's try to integrate over θ first. Starting from (6.8), we get

$$
\int \left[\prod_{i=1}^{n} \frac{e^{-\lambda_i}\lambda_i^{x_i}}{x_i!} \right] \times \frac{\beta^\alpha}{\Gamma(\alpha)} \underbrace{\left[\int \theta^{n+\alpha-1} e^{-(\beta + \sum_{i=1}^{n}\lambda_i)\theta} \, d\theta \right]}_{\star} d\lambda_1 \ldots d\lambda_n
$$

$$
= \int \left[\prod_{i=1}^{n} \frac{e^{-\lambda_i}\lambda_i^{x_i}}{x_i!} \right] \times \frac{\beta^\alpha}{\Gamma(\alpha)} \times \frac{\Gamma(n+\alpha)}{(\beta + \sum_{i=1}^{n}\lambda_i)^{n+\alpha}} \, d\lambda_1 \ldots d\lambda_n
$$

$$
= \frac{(\beta^\alpha)\Gamma(n+\alpha)}{\Gamma(\alpha) \prod_{i=1}^{n} x_i!} \underbrace{\int \frac{\left(e^{-\sum_{i=1}^{n}\lambda_i}\right)\left(\prod_{i=1}^{n}\lambda_i^{x_i}\right)}{(\beta + \sum_{i=1}^{n}\lambda_i)^{n+\alpha}} \, d\lambda_1 \ldots d\lambda_n}_{?}, \quad (6.9)
$$

where the integral marked by "★" can be computed in much the same way as we have always done so far in this book—that is, by exploiting the fact that any density function must integrate to one (Exercise 6.3a). But this last (n-dimensional) integral (marked by "?") over $\lambda_1, \ldots, \lambda_n$ looks formidably impossible due to the denominator, $(\beta + \sum_{i=1}^{n} \lambda_i)^{n+\alpha}$.

Alternatively, we can try to integrate over $\lambda_1, \ldots, \lambda_n$ first. Again, starting from (6.8), we get

$$\int \left\{ \prod_{i=1}^{n} \left[\underbrace{\int \frac{e^{-\lambda_i} \lambda_i^{x_i}}{x_i!} \times \theta e^{-\theta \lambda_i} \, d\lambda_i}_{\bigstar\bigstar} \right] \right\} \times \left[\frac{\beta^\alpha}{\Gamma(\alpha)} \theta^{\alpha-1} e^{-\beta\theta} \right] d\theta$$

$$= \int \left[\frac{\theta}{1+\theta} \right]^n \times \left[\frac{1}{1+\theta} \right]^{n\bar{x}} \times \left[\frac{\beta^\alpha}{\Gamma(\alpha)} \theta^{\alpha-1} e^{-\beta\theta} \right] d\theta$$

$$= \frac{\beta^\alpha}{\Gamma(\alpha)} \underbrace{\int \frac{\theta^{n+\alpha-1} e^{-\beta\theta}}{(1+\theta)^{n(\bar{x}+1)}} \, d\theta,}_{??} \tag{6.10}$$

where the integral marked by "★★" is actually identical to the one leading to Equation (6.1) (see also Exercise 6.3b). But, again, we find ourselves unable to evaluate this last integral (marked by "??") over θ due to the denominator, $(1+\theta)^{n(\bar{x}+1)}$, even though it is "merely" a one-dimensional integral this time.

In fact, the full posterior here is simply intractable, and this is typical of most Bayesian problems. The only currently available solution is that we can use various Monte Carlo methods to draw a sample from the posterior, say,

$$(\lambda_1^{(b)}, \ldots, \lambda_n^{(b)}, \theta^{(b)}) \sim \pi(\lambda_1, \ldots, \lambda_n, \theta | x_1, x_2, \ldots, x_n), \quad b = 1, 2, \ldots, B,$$

which will then allow us to approximate various quantities such as the posterior expectation,

$$\mathbb{E}(\lambda_1, \ldots, \lambda_n, \theta | X_1 = x_1, \ldots, X_n = x_n) \approx \frac{1}{B} \sum_{b=1}^{B} (\lambda_1^{(b)}, \ldots, \lambda_n^{(b)}, \theta^{(b)}),$$

numerically (see also Remark 5.1 in section 5.1).

A particular algorithm that we can use here is the Gibbs sampler (see Mathematical Insert 10 above), which generates Monte Carlo samples following a multidimensional distribution by sequentially sampling from the conditional distribution of each coordinate given the rest, one coordinate after another. The integral marked by "★★" leading to Equation (6.10) implies (Exercise 6.3b) that

$$\pi(\lambda_i|\lambda_1,\ldots,\lambda_{i-1},\lambda_{i+1},\ldots,\lambda_n,\theta,x_1,\ldots,x_n) \sim \text{Gamma}(x_i+1,1+\theta), \quad (6.11)$$

for each $i = 1,2,\ldots,n$ and the one marked by "★" leading to Equation (6.9) implies (Exercise 6.3a) that

$$\pi(\theta|\lambda_1,\ldots,\lambda_n,x_1,\ldots,x_n) \sim \text{Gamma}\left(\alpha+n,\beta+\sum_{i=1}^{n}\lambda_i\right). \quad (6.12)$$

The Gibbs sampler simply cycles through (6.11)–(6.12) and draws a long sequence of samples,

$$\lambda_1^{(1)},\ldots,\lambda_n^{(1)},\theta^{(1)},\lambda_1^{(2)},\ldots,\lambda_n^{(2)},\theta^{(2)},\ldots,$$

as outlined in Table 6.2. Just like in Example 6.2, here, we see again that our knowledge of each λ_i is informed not only by X_i alone but also by θ, which is informed by all of $\lambda_1,\ldots,\lambda_n$, and hence by all of $X_1,\ldots,X_n$ as well.

Figure 6.1 shows some results for the data given in Table 6.1. We can see that the answers produced by the hierarchical and empirical Bayes approaches about $\lambda_1,\ldots,\lambda_n$ and θ are essentially indistinguishable from each other— except that, for θ, the empirical Bayes approach gives only a numeric estimate $\hat{\theta}$ rather than a posterior distribution $\pi(\theta|x_1,\ldots,x_n)$, but the two answers are still in agreement in the sense that $\hat{\theta}$ is near the center of $\pi(\theta|x_1,\ldots,x_n)$. □

Table 6.2 The Gibbs sampler for Example 6.3

---

initialize $t = 1$ and $\lambda_i^{(0)} = x_i$ for $i = 1,2,\ldots,n$
while $(t < t_{\max})$
    draw $\theta^{(t)} \sim \text{Gamma}(\alpha+n,\beta+\sum_{i=1}^{n}\lambda_i^{(t-1)})$
    draw $\lambda_i^{(t)} \sim \text{Gamma}(x_i+1,\theta^{(t)}+1)$ for $i = 1,2,\ldots,n$
    increment $t = t+1$
end while
return $\{(\lambda_1^{(t)},\ldots,\lambda_n^{(t)},\theta^{(t)}) : t > t_{\text{burn.in}}\}$

Note: Figure 6.1 uses $\alpha = \beta = 0.001$, $t_{\max} = 20,000$, and $t_{\text{burn.in}} = 15,000$.

Source: authors.

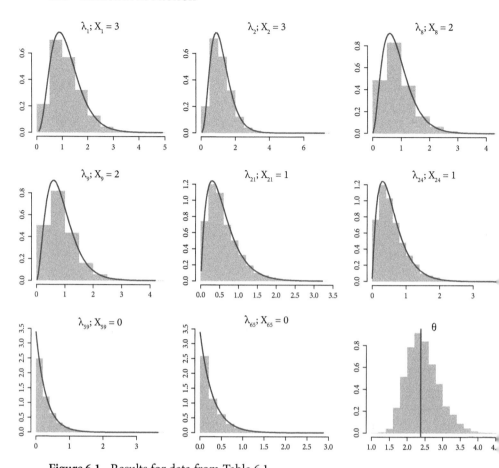

Figure 6.1 Results for data from Table 6.1.

Note: Histograms are posterior samples of λ_i, for a few selected $i \in \{1, 2, \ldots, 100\}$, and of θ. Solid lines are empirical Bayes posterior densities $\pi(\lambda_i | x_i; \widehat{\theta})$ for each corresponding λ_i, and the corresponding empirical Bayes estimate $\widehat{\theta}$ for θ.

Source: authors.

Remark 6.3. In the preceding example, there are only two levels of hierarchy, as shown below.

$$(\alpha, \beta) \quad \rightarrow \quad \theta \quad \rightarrow \quad \begin{cases} \lambda_1 & \rightarrow & X_1 \\ \lambda_2 & \rightarrow & X_2 \\ & \vdots & \\ \lambda_n & \rightarrow & X_n \end{cases}$$

$$\underbrace{}_{\substack{\text{Level 2} \\ \text{Hyperprior}}} \quad \underbrace{}_{\substack{\text{Level 1} \\ \text{Prior}}} \quad \underbrace{\phantom{\text{Level 0 Model}}}_{\substack{\text{Level 0} \\ \text{Model}}}$$

While the hyperparameter (θ) at the first level is treated as a random variable, those at the second—and here, final—level (α and β) must be treated as fixed constants in order for the method to work in practice. As such, their values must be specified. One way to do this is to choose $\alpha = \beta = \varepsilon$ for some tiny $\varepsilon > 0$. The argument is that such a prior will have negligible influence on the outcome of the analysis, as can be seen from Equations (6.2) and (6.12). Such priors are called *non-informative priors*.[2] □

Remark 6.4. Example 6.3 also reveals that, in order to use the Gibbs sampler, conjugate priors are usually required because this would allow us to use our favorite integration technique (see Example 3.4 in section 3.1) to derive the various conditional distributions needed.[3] But if non-conjugate priors are used, then more general-purpose sampling algorithms will usually be needed—for example, the Metropolis–Hastings algorithm (see Appendix 6.A). □

Exercise 6.3. Continue Example 6.3 by filling in various details.

(a) Complete the steps leading to Equation (6.9) by evaluating the integral marked "★", and prove (6.12).
(b) Complete the steps leading to Equation (6.10) by evaluating the integral marked "★★", and prove (6.11).
(c) Implement the Gibbs sampler outlined in Table 6.2, apply it to the data set in Table 6.1 (section 6.2), and obtain a sample from the posterior distribution, $\pi(\lambda_1, \ldots, \lambda_n, \theta | x_1, \ldots, x_n)$.
(d) Reproduce Figure 6.1 with your posterior sample from (c).

[*Think: In Table 6.2, why is it reasonable to initialize each λ_i with x_i?*] □

Exercise 6.4. Table 6.3 shows the number of disasters at coal mines during the period of 1851–1962. It is believed that, at some point, things got better and the average number of disasters per year dropped significantly. Let X_t be

[2] In fact, here, one can simply choose $\alpha = \beta = 0$ as well, except, in this case, the corresponding prior, Gamma(0, 0), is ill-defined and not a proper distribution; it is called an *improper prior*, but the resulting posterior is still a perfectly proper distribution. Many non-informative priors are indeed improper priors.
[3] Of course, that the Gibbs sampler can be implemented doesn't mean it will necessarily always be efficient. Sometimes, it can be painfully slow to reach "burn in"; at other times, it is difficult even to verify whether the sampler has reached "burn in" at all. While we have come to know a lot about the Gibbs sampler over the years, implementing it is still very much a skillful art.

Table 6.3 Disasters at coal mines during the period 1851–1962

	i									
Year	0	1	2	3	4	5	6	7	8	9
$1851+i$	4	5	4	1	0	4	3	4	0	6
$1861+i$	3	3	4	0	2	6	3	3	5	4
$1871+i$	5	3	1	4	4	1	5	5	3	4
$1881+i$	2	5	2	2	3	4	2	1	3	2
$1891+i$	2	1	1	1	1	3	0	0	1	0
$1901+i$	1	1	0	0	3	1	0	3	2	2
$1911+i$	0	1	1	1	0	1	0	1	0	0
$1921+i$	0	2	1	0	0	0	1	1	0	2
$1931+i$	3	3	1	1	2	1	1	1	1	2
$1941+i$	4	2	0	0	0	1	4	0	0	0
$1951+i$	1	0	0	0	0	0	1	0	0	1
$1961+i$	0	1								

Source: authors.

the number of disasters in year t, for $t = 1, 2, \ldots, n$, where $t = 1$ corresponds to the year of 1851 and $t = n$, the year of 1962. We will consider the following model:

$$X_t \sim \begin{cases} \text{Poisson}(\lambda_1), & \text{for} \quad t = 1, 2, \ldots, \theta; \\ \text{Poisson}(\lambda_2), & \text{for} \quad t = \theta + 1, \theta + 2, \ldots, n. \end{cases}$$

Moreover, we will take a Bayesian approach and further assume that the prior distributions of $\lambda_1, \lambda_2, \theta$ are independent; $\lambda_i \sim \text{Gamma}(0.05, 0.01)$ for $i = 1, 2$; and $\theta \sim \text{Uniform}\{1, 2, \ldots, n\}$.

(a) Show that

$$\pi(\theta|\lambda_1, \lambda_2, x_1, \ldots, x_n) = \frac{A_\theta B_\theta}{\sum_{\theta'=1}^{n} A_{\theta'} B_{\theta'}}, \quad \text{for} \quad \theta = 1, 2, \ldots, n,$$

where

$$A_\theta = \prod_{t=1}^{\theta} \frac{e^{-\lambda_1}\lambda_1^{x_t}}{x_t!} \quad \text{and} \quad B_\theta = \begin{cases} \displaystyle\prod_{t=\theta+1}^{n} \frac{e^{-\lambda_2}\lambda_2^{x_t}}{x_t!}, & \theta < n; \\ 1, & \theta = n. \end{cases}$$

(b) Find $\pi(\lambda_1|\lambda_2, \theta, x_1, \ldots, x_n)$ and $\pi(\lambda_2|\lambda_1, \theta, x_1, \ldots, x_n)$.

(c) Implement a Gibbs sampler to obtain a sample from the posterior distribution, $\pi(\lambda_1, \lambda_2, \theta | x_1, \ldots, x_n)$. Make a marginal histogram for λ_1, λ_2, and θ, respectively.

(d) When did the mining disasters situation start to improve? Using the posterior sample, compute a numeric estimate of

$$\mathbb{E}\left(\frac{\lambda_2}{\lambda_1}\middle| X_1, \ldots, X_n\right) \quad \text{and} \quad \sqrt{\mathbb{V}\mathrm{ar}\left(\frac{\lambda_2}{\lambda_1}\middle| X_1, \ldots, X_n\right)},$$

respectively. How would you quantify the improvement?

[*Hint: It is numerically better to compute* $\exp\left[\sum \log(a_t)\right]$ *rather than* $\prod a_t$, *especially for* $a_t \in (0, 1)$. *(Why?)*][4] □

Appendix 6.A General sampling algorithms

The Gibbs sampler is a special case of a more general sampling algorithm.

6.A.1 Metropolis algorithm

Suppose we'd like to draw samples from a target distribution, $f(z)$. The Metropolis algorithm proceeds as follows. Start with an initial guess z_0. At each iteration t, let $z = z_{t-1}$ denote the current point. First, the algorithm makes a proposal based on z, that is,

$$z' \sim T(z, z'),$$

where $T(u, v) = T(v, u)$ is a symmetric proposal distribution, in which the first argument z is fixed and $\int T(z, z') dz' = 1$. For example, $z' \sim N(z, \sigma^2 I)$.[5] Then, with probability

$$\min\left\{\frac{f(z')}{f(z)}, 1\right\}, \tag{6.13}$$

the algorithm accepts the proposal and $z_t = z'$; otherwise, it rejects the proposal and $z_t = z_{t-1} = z$. After "burn in", a sample $\{z_t : t > t_{\mathrm{burn.in}}\}$ from the target distribution $f(\cdot)$ is obtained.

---

[4] This exercise has been adapted from the text by G. H. Givens and J. A. Hoeting [12].
[5] According to this distribution, points closer to z will have a higher chance of being proposed than those farther away.

It is easy to appreciate why this algorithm is so "perfect" for Bayesian statistics. In the Bayesian context, the target distribution is usually a (multi-dimensional) posterior distribution of some sort,

$$\pi(\boldsymbol{\theta}|x_1,\ldots,x_n) = \frac{f(x_1,\ldots,x_n|\boldsymbol{\theta}) \cdot \pi(\boldsymbol{\theta})}{m(x_1,\ldots,x_n)}, \quad \boldsymbol{\theta} \in \mathbb{R}^d.$$

The numerator above is easy to evaluate at any $\boldsymbol{\theta}$; the main bottleneck is the denominator,

$$m(x_1,\ldots,x_n) = \int f(x_1,\ldots,x_n|\boldsymbol{\theta}) \cdot \pi(\boldsymbol{\theta})\, d\boldsymbol{\theta},$$

but it is *not* needed when evaluating the Metropolis ratio,

$$\frac{\pi(\boldsymbol{\theta}'|x_1,\ldots,x_n)}{\pi(\boldsymbol{\theta}\,|x_1,\ldots,x_n)}.$$

It is thus a very general algorithm for obtaining a posterior sample, even when $\pi(\boldsymbol{\theta})$ is not chosen to be conjugate to $f(x_1,\ldots,x_n|\boldsymbol{\theta})$.

6.A.2 Some theory

The reason why the algorithm works is because the target distribution $f(\cdot)$ is easily shown to satisfy a key condition which makes it the stationary distribution of the Markov chain[6] produced by the algorithm.

Definition 7 (Stationary distribution). *For a Markov chain with transition function $A(\boldsymbol{u}, \boldsymbol{v})$, a distribution $\pi(\cdot)$ that satisfies*

$$\int \pi(\boldsymbol{u})A(\boldsymbol{u}, \boldsymbol{v})d\boldsymbol{u} = \pi(\boldsymbol{v}) \qquad (6.14)$$

is called its stationary distribution. ☐

The meanings of "transition function" and "stationary distribution" are best explained with a simple example.

---

[6] Like the Poisson process (see section 4.3.2 and Appendix 4.B), the Markov chain is another special *stochastic process*, which we do not really study in this book. In a nut shell, a (discrete-time) Markov chain can be in one of many different states at any given time, and there is a transition function, say, $A(u, v)$, that describes the relative likelihood that it will end up in state v at the next time point, given that it is currently in state u, for all combinations of (u, v).

Example 6.4. Consider a simple, two-state, Markov chain, that is, $u, v \in \{1, 2\}$. Going from time $t - 1$ to time t, we have

$$\underbrace{\mathbb{P}(\text{in state 1 at time } t)}_{\mathbb{P}_t(1)}$$

$$= \underbrace{[\mathbb{P}(\text{in state 1 at time } t - 1)]}_{\mathbb{P}_{t-1}(1)} \underbrace{[\mathbb{P}(\text{go from state 1 to state 1})]}_{A(1,1)}$$

$$+ \underbrace{[\mathbb{P}(\text{in state 2 at time } t - 1)]}_{\mathbb{P}_{t-1}(2)} \underbrace{[\mathbb{P}(\text{go from state 2 to state 1})]}_{A(2,1)}$$

by the law of total probability (Chapter 2). A similar equation can be written for the probability of being "in state 2 at time t", and we can write these two equations compactly in matrix form, as follows:

$$\begin{bmatrix} \mathbb{P}_{t-1}(1) & \mathbb{P}_{t-1}(2) \end{bmatrix} \begin{bmatrix} A(1,1) & A(1,2) \\ A(2,1) & A(2,2) \end{bmatrix} = \begin{bmatrix} \mathbb{P}_t(1) & \mathbb{P}_t(2) \end{bmatrix}.$$

Thus, if the state space is discrete, we can always write

$$\sum_u \mathbb{P}_{t-1}(u)A(u, v) = \mathbb{P}_t(v),$$

and "stationary" merely means that the distribution (over the entire state space) at time t is the same as the one at time $t - 1$, that is, $\mathbb{P}_{t-1}(\cdot) = \mathbb{P}_t(\cdot)$. Equation (6.14) is the equivalent statement for continuous state spaces. □

A key condition for a distribution $f(\cdot)$ to be the stationary distribution of a Markov chain with transition function $A(u, v)$ is if

$$f(u)A(u, v) = f(v)A(v, u), \quad \text{for any } u, v. \tag{6.15}$$

To see this, notice that, if $f(\cdot)$ satisfies Equation (6.15) above, then

$$\int f(u)A(u, v)du = \int f(v)A(v, u)du = f(v) \underbrace{\int A(v, u)du}_{\overset{(\star)}{=} 1} = f(v),$$

which shows that $f(\cdot)$ is a stationary distribution by Definition 7. [*Think: Why is the step labelled "($\star$)" true?*]

For the Metropolis algorithm, the actual transition function of the underlying Markov chain is:

$$A(z, z') = \underbrace{T(z, z')}_{\text{Proposal}} \times \underbrace{\min\left\{\frac{f(z')}{f(z)}, 1\right\}}_{\text{Acceptance}}.$$

So we get

$$
\begin{aligned}
f(z)A(z, z') &= \min\left\{f(z)T(z, z')\frac{f(z')}{f(z)}, f(z)T(z, z')\right\} \\
&= \min\{f(z')T(z, z'), f(z)T(z, z')\} \\
&= \min\{f(z')T(z', z), f(z)T(z, z')\} \quad\quad (6.16)
\end{aligned}
$$

since $T(\cdot, \cdot)$ is symmetric. But the final expression in Equation (6.16) is symmetric with respect to z and z', so

$$f(z)A(z, z') = f(z')A(z', z),$$

which means $f(\cdot)$ satisfies Equation (6.15) and is thus a stationary distribution of the Markov chain generated by the sequential algorithm.

Remark 6.5. One can be misled easily by the analysis above to believe that the Metropolis algorithm "should just work" for any symmetric proposal distribution. While this is true in principle, in practice there is a big difference between a Markov chain that *eventually* reaches its stationary distribution, and one that reaches its stationary distribution *in a reasonable number of steps*. In this regard, the choice of the proposal distribution is critical, and a successful implementation of the Metropolis algorithm is still very much a skillful art; see Exercise 6.5 below. ☐

Exercise 6.5. Implement a Metropolis algorithm to draw samples from $N(0, 1)$ using the simple proposal distribution $x_t \sim N(x_{t-1}, \sigma^2)$. Try $\sigma = 0.1$, 2.38, and 25. Which σ appears to work best? [*Note: Of course, nobody really uses the Metropolis algorithm to draw samples from $N(0, 1)$; there are much better algorithms for doing so. This is merely a toy exercise so that one can appreciate the fact that proposal distributions matter in practice. The mysterious "2.38" is the result of some rather brilliant theoretical analysis [13], which we won't explain here.*] ☐

6.A.3 Metropolis–Hastings algorithm

If the proposal distribution $T(u, v)$ is not symmetric, one simply has to modify the acceptance probability to be

$$\min\left\{\frac{f(z')T(z',z)}{f(z)T(z,z')}, 1\right\}. \tag{6.17}$$

The resulting algorithm is known as the Metropolis–Hastings algorithm.

Exercise 6.6. Show that

$$f(z) \times T(z, z') \times \min\left\{\frac{f(z')T(z',z)}{f(z)T(z,z')}, 1\right\}$$

is symmetric with respect to z and z', and hence conclude that $f(\cdot)$ is a stationary distribution of the Markov chain generated by the Metropolis–Hastings algorithm. □

The Gibbs sampler (Mathematical Insert 10 in section 6.3) can be viewed as a special Metropolis–Hastings algorithm. At any given step with current sample $z = (z_1, z_2, \ldots, z_d)^\mathsf{T}$, the algorithm simply makes a very special proposal,

$$z' = (z_1, \ldots, z_{j-1}, z'_j, z_{j+1}, \ldots, z_d)^\mathsf{T}$$

for some $j \in \{1, 2, \ldots, d\}$. The only new element/coordinate, z'_j, is proposed (here, drawn) from the conditional distribution

$$z'_j \sim f(z_j|z_{-j}) \equiv f(z_j|z_1, \ldots, z_{j-1}, z_{j+1}, \ldots z_d),$$

whereas all other elements are kept the same, that is, $z'_k = z_k$ for all $k \neq j$ or simply $z'_{-j} = z_{-j}$.

Exercise 6.7. Show that, with $T(z, z') = f(z'_j|z_{-j})$ and $z'_{-j} = z_{-j}$, the acceptance probability given by Equation (6.17) is equal to one. [*Note: This means the special proposal made by the Gibbs sampler at any given step is always accepted under the Metropolis–Hastings principle.*] □

PART III
FACING UNCERTAINTY

Synopsis: Although one usually does not care much about parameters that don't have intrinsic scientific meaning, for those that do, it is important to explicitly describe how much uncertainty we have about them and to take that into account when making decisions.

7

Interval Estimation

Before we move on to the last part of this book, here is a quick recapitulation of Part II. In statistics, we would like to say something about the parameter, θ, of an underlying probability model based on the random variables, $X_1, X_2, \ldots, X_n$, that it generates. In the frequentist approach, θ is assumed to be fixed (albeit unknown), and the main objective is to find an estimator $\hat{\theta} = g(X_1, X_2, \ldots, X_n)$ in order to estimate it; whereas, in the Bayesian approach, θ is assumed to be a random variable itself, and the main objective is to find its posterior distribution $\pi(\theta | x_1, x_2, \ldots, x_n)$.

Sometimes, especially in modern applications of artificial intelligence, the purpose of the model is largely utilitarian. We don't necessarily believe it's even close to being the true data-generating process but concede, nonetheless, that it still describes the process reasonably well and that we may even rely on it to make certain predictions.

For example, according to the model $f(x; \theta)$, the event "$X \in A$" will happen with probability

$$\mathbb{P}(X \in A) = \int_{x \in A} f(x; \theta)dx,$$

which we cannot compute if we do not know the parameter θ. Having first estimated the parameter θ, however, a frequentist can now predict that it will happen with probability

$$\hat{\mathbb{P}}(X \in A) \equiv \int_{x \in A} f(x; \hat{\theta})dx, \tag{7.1}$$

whereas a Bayesian, having computed the posterior distribution of $\theta | X_1, \ldots, X_n$, may tell us that it will happen with probability

$$
\begin{aligned}
\mathbb{P}(X \in A | X_1, \ldots, X_n) &= \int \mathbb{P}(X \in A | \theta)\pi(\theta | x_1, \ldots, x_n)d\theta \\
&= \int \left[\int_{x \in A} f(x | \theta)dx \right] \pi(\theta | x_1, \ldots, x_n)d\theta. \tag{7.2}
\end{aligned}
$$

Essential Statistics for Data Science. Mu Zhu, Oxford University Press. © Mu Zhu (2023).
DOI: 10.1093/oso/9780192867735.003.0007

In these situations, we don't necessarily care too much about the parameter θ itself, but we still must say something about it because otherwise we cannot actually make use of the model.

Remark 7.1. Equation (7.2) also highlights the fact that, even if there is a closed-form analytic expression for the posterior distribution $\pi(\theta|x_1, \ldots, x_n)$ itself, predictive quantities such as (7.2) are often still intractable. This explains why, in practice, it is more important to be able to sample from the posterior distribution (e.g. Mathematical Insert 10 in section 6.3) than to derive an analytic expression for it. [*Think: How would you compute/approximate (7.2) with a sample from the posterior?*]   □

Other times, especially in scientific applications, we may take the model more seriously, and the parameter θ itself will often have some special meaning, too.

Example 7.1.   Consider the model given in Exercise 5.10 (section 5.3). As a concrete example, X_i here may be a binary indicator of whether individual i is obese, defined perhaps as having a body mass index (BMI) of over 30, and v_i may be a covariate, such as the average number of hours this individual sleeps per night minus eight. The reason for the "minus eight" is so that the quantity, $\exp(\alpha)/[1+\exp(\alpha)]$, can be interpreted conveniently as the probability of being obese for someone who sleeps an average of eight hours per night. [*Think: Why?*] This quantity (and hence the parameter α itself) is not really of any intrinsic interest; it merely provides a kind of baseline value for the population under study.

By contrast, the other parameter β is more interesting. Equation (5.13) in section 5.3 specifies that the *odds*—here, of an individual i being obese—is equal to

$$\frac{p_i}{1 - p_i} = \exp(\alpha + \beta v_i),$$

from which it is then easy to derive that an extra hour of sleep (i.e. going from v_i to $v_i + 1$) will change the odds by a factor of $\exp(\beta)$. (Try it.) Therefore, if $\beta = 0$, it may suggest that sleep has little effect on obesity; whereas, if $\beta = -0.1$, the implication is that every extra hour of sleep will approximately reduce the odds of being obese by about 10% ($e^{-0.1} \approx 0.9$). So, scientists may very well be interested in the value of β itself, even if the model itself may be overly simplistic.   □

In these situations, how much uncertainty we have about θ becomes an important question. In Example 7.1 above, we may estimate the parameter β to be -0.1 and still be uncertain enough about its actual value that we cannot confidently rule out the possibility that it may be equal to zero after all. This is no light matter; a correct or incorrect understanding of the relationship between sleep and obesity undoubtedly can affect many human lives.

Of course, even if the parameter θ may not have any meaning and we may not care much about it, our uncertainty about its value is still important. For example, having estimated θ, we can always compute Equation (7.1) but, if there is a lot of uncertainty about what θ really is, then we may not want to trust the answer as much.

7.1 Uncertainty quantification

One way to quantify the uncertainty about θ is by coming up with probability statements such as

$$\mathbb{P}(1.913 < \theta < 2.004) = 95\%. \tag{7.3}$$

7.1.1 Bayesian version

For a Bayesian, the meaning of such a statement is straightforward since the parameter θ is treated as a random variable, so one can indeed make various probability statements about it. For any given $0 < \alpha \ll 1$, one simply uses the posterior distribution $\pi(\theta | x_1, \ldots, x_n)$ to identify threshold values c_α^{lo} and c_α^{hi} so that

$$\mathbb{P}(c_\alpha^{\text{lo}} < \theta < c_\alpha^{\text{hi}} | X_1, \ldots, X_n) = 1 - \alpha.$$

These choices are not unique. A relatively easy-to-implement strategy in practice is to choose c_α^{lo} and c_α^{hi} such that

$$\int_{-\infty}^{c_\alpha^{\text{lo}}} \pi(\theta | x_1, \ldots, x_n) d\theta = \int_{c_\alpha^{\text{hi}}}^{\infty} \pi(\theta | x_1, \ldots, x_n) d\theta = \frac{\alpha}{2};$$

that is, just choose c_α^{lo} to be the $\alpha/2 \times 100\%$ quantile of the posterior distribution and c_α^{hi} to be the $(1 - \alpha/2) \times 100\%$ quantile.

7.1.2 Frequentist version

For a frequentist, however, the meaning of a statement like (7.3) can be quite puzzling since the parameter θ is treated as a non-random—albeit unknown— constant, so what does it mean to make a probability statement about something that is not even random? It is best to illustrate the gist of the underlying logic with a concrete toy example.

Example 7.2. Suppose $X_1, X_2, \ldots, X_n \overset{iid}{\sim} N(\mu, \sigma^2)$. We will first assume that the parameter σ^2 is a known constant so that there is just one unknown parameter to deal with; this (unrealistic) assumption will be removed later in Example 7.3. We can easily estimate μ with $\hat{\mu} = \bar{X}$. Then, surely the amount of uncertainty we have about μ must be contained in the distribution of the estimator $\hat{\mu} = \bar{X} \sim N(\mu, \sigma^2/n)$. Where else can such information come from?! We can now make a series of logical deductions:

$$\bar{X} \sim N(\mu, \sigma^2/n) \overset{(a)}{\Rightarrow} \bar{X} - \mu \sim N(0, \sigma^2/n)$$

$$\overset{(b)}{\Rightarrow} \mathbb{P}\left(\frac{-1.96\sigma}{\sqrt{n}} < \bar{X} - \mu < \frac{1.96\sigma}{\sqrt{n}}\right) = 95\%$$

$$\overset{(c)}{\Rightarrow} \mathbb{P}\left(\bar{X} - \frac{1.96\sigma}{\sqrt{n}} < \mu < \bar{X} + \frac{1.96\sigma}{\sqrt{n}}\right) = 95\%.$$

The first step (a) simply applies a linear transformation to the random variable $\bar{X}$; the second step (b) is based on taking the 2.5% and 97.5% quantiles of the corresponding normal distribution (see Fun Box 3 below); the last step (c) merely rearranges the inequalities so that μ alone is sandwiched in the middle; and we have produced a statement in the form of (7.3).   □

It is clear from Example 7.2 that the probability statement is really about the estimator $\hat{\mu} = \bar{X}$, *not* about the parameter μ, so there is no logical contradiction for frequentists after all because the estimator *is* a random variable and *does* have a distribution; see section 4.1. Nevertheless, after the rearrangement operation in step (c), a probability statement about $\hat{\mu}$ can be used to quantify our uncertainty about μ.

For precisely this reason, frequentist statisticians insist that, when describing (7.3), it is wrong to say, "there is a 95% probability that the parameter θ will *fall into* the interval (1.913, 2.004)"; rather, one should say, "there is a 95% probability that the interval (1.913, 2.004) will *contain* the parameter θ".

Fun Box 3

The two-standard-deviation rule of thumb. The 2.5% and 97.5% quantiles of the standard N(0, 1) normal distribution are −1.96 and +1.96, respectively. These can be shifted (by the mean) and scaled (by the standard deviation) to obtain the corresponding quantiles of any $N(\mu, \sigma^2)$. In practice, one often rounds up $1.96 \approx 2$, leading to the widely quoted "two-standard-deviation rule of thumb", which says there is $\approx$ 95% chance that a *normally distributed* random quantity will fall between two standard deviations of the mean (also see Figure 2.4 in section 2.4):

$$\mathbb{P}(|X - \mu| < 2\sigma) \approx 95\%.$$

Without normality, we can conclude in general—by applying Chebyshev's inequality (see Exercise 2.6 section 2.3.2) with $k = 2$—that there is $\geq$ 75% chance that a random quantity will fall between two standard deviations of the mean:

$$\mathbb{P}(|X - \mu| < 2\sigma) = 1 - \mathbb{P}(|X - \mu| \geq 2\sigma) \geq 1 - \frac{1}{2^2} = \frac{3}{4} = 75\%.$$

The normal distribution is a lot tighter than that. The human IQ is famously distributed as $N(100, 15^2)$, so we know that only about 5% of the population have IQs outside the (70, 130) range, and you can claim to be smarter than 97.5% of the people on the planet if you have an IQ of above 130.

The parameter is not random but the interval is—if we had a different data set, we would obtain a slightly different interval.

7.2 Main difficulty

The process of coming up with statements like Equation (7.3) is sometimes referred to as *interval estimation*. The resulting interval is referred to as a *confidence interval*, if it is constructed with a frequentist approach, or as a *credible interval* if it is constructed with a Bayesian approach, so as to differentiate the two. As we have explained in section 7.1, although their practical utilities are similar, they are fundamentally very different creatures! Because the Bayesian credible interval is conceptually more straightforward and the frequentist confidence interval more intricate, we will focus on the latter for the remainder of this chapter.

The gist of the logic in Example 7.2 is that, if we know the distribution of $\hat{\theta}$, or rather that of $\hat{\theta} - \theta$, we can simply choose $(c_\alpha^{lo}, c_\alpha^{hi})$—for example, the $\alpha/2 \times 100\%$ and $(1 - \alpha/2) \times 100\%$ quantiles of that distribution—so that

$$\mathbb{P}(c_\alpha^{lo} < \hat{\theta} - \theta < c_\alpha^{hi}) = 1 - \alpha.$$

Then, a simple rearrangement of the inequalities as step (c) in Example 7.2 will yield

$$\left\{\theta : c_\alpha^{lo} < \hat{\theta} - \theta < c_\alpha^{hi}\right\} = (\hat{\theta} - c_\alpha^{hi}, \hat{\theta} - c_\alpha^{lo})$$

as a $(1 - \alpha) \times 100\%$ confidence interval for θ. In theory, the logic is not all that complex after all. But in practice, as we have seen in section 3.2.1, it is usually not so easy to work out the distribution of $\hat{\theta}$ and hence that of $\hat{\theta} - \theta$ as well. In fact, even a minor modification to Example 7.2 can make this task substantially harder.

Example 7.3. Suppose $X_1, X_2, \ldots, X_n \overset{iid}{\sim} N(\mu, \sigma^2)$. But now, we will assume that both μ an σ^2 are unknown parameters. The situation is thus more realistic than in Example 7.2, but the model is admittedly still a very simple one. Clearly, the logic from steps (a)–(c) in Example 7.2 still applies, except that the final answer is not usable—it contains an unknown parameter σ, so it's not really an answer—and we appear to be stuck; the distribution of $\hat{\mu} = \bar{X}$ will depend on the additional unknown parameter σ^2, whether we like it or not.

Let

$$S^2 \equiv \frac{1}{n-1} \sum_{i=1}^{n} (X_i - \bar{X})^2. \tag{7.4}$$

This quantity is called the *sample variance*. Exercise 5.7 (section 5.2) explains why S^2 has conventionally been the default estimator of σ^2, not $\hat{\sigma}_{mle}^2$. While

$$Z \equiv \frac{\bar{X} - \mu}{\sigma/\sqrt{n}} \sim N(0, 1)$$

follows the standard normal distribution, a celebrated early achievement of statistics is that, by replacing the unknown σ with S, we can actually work out the distribution of

$$T \equiv \frac{\bar{X} - \mu}{S / \sqrt{n}} \tag{7.5}$$

in closed form! The ensuing distribution is now known as Student's t-distribution (or simply the t-distribution) with $(n - 1)$ degrees of freedom, and denoted as $t_{(n-1)}$. We will omit the exact expression of its density function as it is not too relevant, but it is important to appreciate why it was tedious at best to derive such a result.

The sample variance S^2 is a function of the underlying random variables $X_1, \ldots, X_n$, too—in fact, a more complicated one than $\bar{X}$ itself. So the quantity T, with S^2 appearing in the denominator after being "square-rooted", is a rather complicated (and nonlinear) function of $X_1, \ldots, X_n$. By contrast, the quantity Z is simply a linear function of $X_1, \ldots, X_n$. That's why working out the exact distribution of the new random variable T was not a trivial feat. (Don't try this at home!)

This was a "big deal" because it meant we could use the distribution of T—for example, by relying on its quantiles as we have often done hereinabove—to come up with a probability statement like

$$\mathbb{P}(c_\alpha^{lo} < T < c_\alpha^{hi}) = 1 - \alpha \Rightarrow \mathbb{P}\left(c_\alpha^{lo} < \frac{\bar{X} - \mu}{S / \sqrt{n}} < c_\alpha^{hi}\right) = 1 - \alpha$$

$$\Rightarrow \mathbb{P}\left(\bar{X} - c_\alpha^{hi} \frac{S}{\sqrt{n}} < \mu < \bar{X} - c_\alpha^{lo} \frac{S}{\sqrt{n}}\right) = 1 - \alpha,$$

and hence a confidence interval for μ, even if the other parameter σ^2 remains unknown! □

Remark 7.2. The t-distribution is symmetric about zero, so if one takes c_α^{hi} to be the "usual" $(1 - \alpha/2) \times 100\%$ quantile, the corresponding $\alpha/2 \times 100\%$ quantile "on the other side" will simply be $c_\alpha^{lo} = -c_\alpha^{hi}$. At the "usual" 95% confidence level (i.e. $\alpha = 0.05$), the corresponding threshold $c_{0.05}^{hi} \approx 2$, as long as the underlying degree of freedom, $(n - 1)$, is not too small, as shown below.

df	1	2	3	5	10	20	30	∞
$c_{0.05}^{hi}$	12.71	4.30	3.18	2.57	2.23	2.09	2.04	1.96

That is why many back-of-the-envelope calculations simply use

$$\bar{X} - 2\left(\frac{S}{\sqrt{n}}\right) < \mu < \bar{X} + 2\left(\frac{S}{\sqrt{n}}\right) \tag{7.6}$$

as the "generic" 95% confidence interval for μ, irrespective of the exact degrees of freedom. Also note that, since both $\bar{X}$ and S are functions of $X_1, X_2, \ldots, X_n$ and hence random variables themselves, the resulting interval is random in terms of both its location and its width.   □

7.3 Two useful methods

In both Examples 7.2 and 7.3, we ended up relying on a certain quantity, say, $h(\theta; D)$,

(a) which involved only the parameter θ of interest (but no other unknown parameter) and data $D \equiv \{X_1, \ldots, X_n\}$ and
(b) whose distribution was completely known.

In Example 7.2, the quantity was:

$$h(\mu; D) = \bar{X} - \mu \sim N(0, \sigma^2/n),$$

where σ^2 was known; in Example 7.3, it was:

$$h(\mu; D) = \frac{\bar{X} - \mu}{S/\sqrt{n}} \sim t_{n-1}.$$

Property (b) allowed us to identify a set C_α—for example, $(c_\alpha^{lo}, c_\alpha^{hi})$—such that

$$\mathbb{P}[h(\theta; D) \in C_\alpha] = 1 - \alpha. \tag{7.7}$$

Property (a) then allowed us to "rearrange" or "solve" the relationship

$$h(\theta; D) \in C_\alpha \Rightarrow \theta \in h^{-1}(C_\alpha; D) \equiv \{\theta : h(\theta; D) \in C_\alpha\} \tag{7.8}$$

for θ. But it's not easy to come up with quantities that satisfy both (a) and (b); in statistics, they are called *pivotal quantities*. The choice $\bar{X} - \mu$ seemed easy in Example 7.2 but only because we made an unrealistic additional assumption that the variance parameter σ^2 was known. As soon as such an assumption was removed, coming up with a pivotal quantity required considerable technical prowess.

Exercise 7.1. Suppose $X_1, \ldots, X_n \overset{iid}{\sim} \text{Uniform}(0, \theta)$.

(a) Find the cumulative distribution function of $Y \equiv \max\{X_1, \ldots, X_n\}/\theta$.
(b) Use Y as a pivotal quantity to find a 90% confidence interval for θ.

[*Think: Is there anything "strange" about the resulting confidence interval? See Exercise 5.8 in section 5.2 for the MLE of θ in this case.*]   □

7.3.1 Likelihood ratio

One fairly "generic" pivotal quantity is

$$h(\theta; D) = 2\log\left[\frac{L(\hat{\theta}_{mle}; D)}{L(\theta; D)}\right] = 2[\ell(\hat{\theta}_{mle}; D) - \ell(\theta; D)],$$

where $L(\cdot)$ is the likelihood function and $\ell(\cdot)$ the log-likelihood function (see section 5.1). Remark 7.4 below will further clarify why it is not truly "generic". Often, we will also simply write $L(\cdot)$ and $\ell(\cdot)$ instead of $L(\cdot; D)$ and $\ell(\cdot; D)$— unless their dependence on D requires explicit emphasis in the context (e.g. Exercise 7.5 below).

The quantity inside the square bracket is called the *likelihood ratio*. The logic is quite simple why it can be particularly useful for uncertainty quantification. By definition of the MLE, the likelihood function $L(\theta)$ reaches its maximum at $\theta = \hat{\theta}_{mle}$, but if another candidate parameter θ does not reduce the likelihood function by too much, in the sense that it does not make the aforementioned likelihood ratio too large, then it ought *not* to be excluded as a potentially correct value of the parameter. This logic means we are interested in the set,

$$\left\{\theta : \frac{L(\hat{\theta}_{mle})}{L(\theta)} < c_\alpha\right\}, \tag{7.9}$$

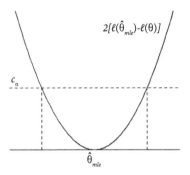

Figure 7.1 The likelihood ratio on the logarithmic scale (or simply the "log-ratio") is, by definition, equal to zero at $\theta = \hat{\theta}_{mle}$. As θ moves away from $\hat{\theta}_{mle}$ in either direction, $\ell(\theta)$ necessarily decreases, so the log-ratio necessarily increases. Stopping this movement when the log-ratio reaches a critical threshold c_α traces out an interval around $\hat{\theta}_{mle}$. The threshold c_α is chosen so that the resulting interval has $(1 - \alpha) \times 100\%$ probability of containing the true parameter value θ.
Source: authors.

for some choice c_α, which puts the probability of the set at exactly $(1 - \alpha) \times 100\%$. Figure 7.1 illustrates this basic idea. But finding the actual threshold c_α requires us to know the distribution of the likelihood ratio itself, and it is usually a complicated function of $D = \{X_1, \ldots, X_n\}$ for sure. That's where the mysterious twice-logarithmic-transform comes in.

Example 7.4. Suppose $X_1, X_2, \ldots, X_n \overset{iid}{\sim} N(\mu, \sigma^2)$. Again, we will first assume that the parameter σ^2 is known. The MLE of μ is simply $\bar{X}$ (see Exercise 5.2; section 5.1.1). Hence, the likelihood ratio is equal to

$$\frac{L(\hat{\mu}_{mle})}{L(\mu)} = \frac{L(\bar{X})}{L(\mu)} = \frac{\prod_{i=1}^{n} \frac{1}{\sqrt{2\pi\sigma^2}} \exp\left[-\frac{(X_i - \bar{X})^2}{2\sigma^2}\right]}{\prod_{i=1}^{n} \frac{1}{\sqrt{2\pi\sigma^2}} \exp\left[-\frac{(X_i - \mu)^2}{2\sigma^2}\right]} = \ldots$$

$$\ldots = \exp\left[\frac{n\bar{X}^2 - 2\mu(n\bar{X}) + n\mu^2}{2\sigma^2}\right] = \exp\left[\frac{n(\bar{X} - \mu)^2}{2\sigma^2}\right], \quad (7.10)$$

where we have omitted a few algebraic steps which do not yield any useful insight by themselves, but readers can try to fill them in on their own. Equation (7.10) now reveals why the twice-logarithmic-transform is so "magical":

$$2 \log \left[\frac{L(\hat{\mu}_{mle})}{L(\mu)} \right] = \frac{n(\bar{X} - \mu)^2}{\sigma^2} = \left[\frac{\bar{X} - \mu}{\sigma / \sqrt{n}} \right]^2. \tag{7.11}$$

The quantity inside the square brackets above is distributed as $N(0, 1)$. This means we know the distribution of the twice-log-transformed likelihood ratio to be $\chi^2_{(1)}$ (see Exercise 3.5 in section 3.2.1).   □

Example 7.4 suggests that, rather than focusing on (7.9), we can equivalently focus on

$$\left\{ \theta : 2[\ell(\hat{\theta}_{mle}) - \ell(\theta)] < c_\alpha \right\} \tag{7.12}$$

instead, since the twice-logarithmic-transform is monotonic, and choose the decision threshold c_α in Equation (7.12) by referring to a chi-squared distribution. This procedure is indeed quite widely applicable, not just to the toy example above. However, to make the recipe fully general, it is necessary to first introduce a further extension.

Often, there will be other parameters (say, ψ) in the model than the one (θ) we are interested in, as in Example 7.3. Under such circumstances, it is necessary to extend the definition of the likelihood ratio to be

$$\Lambda(\theta; \boldsymbol{D}) \equiv \frac{\sup\limits_{\theta', \psi} L(\theta', \psi)}{\sup\limits_{\psi} L(\theta, \psi)} = \frac{L(\hat{\theta}_{mle}, \hat{\psi}_{mle})}{L(\theta, \hat{\psi}_{mle|\theta})}, \tag{7.13}$$

where $\hat{\psi}_{mle|\theta}$ is the constrained (or partial) MLE of ψ while keeping θ fixed.

Example 7.5.   Let's now remove the unrealistic assumption in Example 7.4, which was made merely to facilitate our discussion. With σ^2 also being unknown, we will now have to estimate it twice in order to compute the likelihood ratio (7.13)—once with μ fixed and once without such a constraint. The MLE of σ^2 is

$$\hat{\sigma}^2 = \frac{1}{n} \sum_{i=1}^{n} (X_i - \bar{X})^2;$$

see Exercise 5.2 in section 5.1.1. A similar derivation shows that, if μ is fixed, the corresponding MLE is

$$\widehat{\sigma}_\mu^2 = \frac{1}{n}\sum_{i=1}^n (X_i - \mu)^2.$$

Thus, the likelihood ratio becomes

$$\Lambda(\mu; D) = \frac{L(\widehat{\mu}_{mle}, \widehat{\sigma}_{mle}^2)}{L(\mu, \widehat{\sigma}_{mle|\mu}^2)} = \frac{L(\bar{X}, \widehat{\sigma}^2)}{L(\mu, \widehat{\sigma}_\mu^2)}$$

$$= \frac{\prod_{i=1}^n \frac{1}{\sqrt{2\pi\widehat{\sigma}^2}} \exp\left[-\frac{(X_i - \bar{X})^2}{2\widehat{\sigma}^2}\right]}{\prod_{i=1}^n \frac{1}{\sqrt{2\pi\widehat{\sigma}_\mu^2}} \exp\left[-\frac{(X_i - \mu)^2}{2\widehat{\sigma}_\mu^2}\right]} = \left[\frac{\widehat{\sigma}_\mu^2}{\widehat{\sigma}^2}\right]^{n/2} \times \frac{\exp\left[-\frac{\sum(X_i - \bar{X})^2}{2\widehat{\sigma}^2}\right]}{\exp\left[-\frac{\sum(X_i - \mu)^2}{2\widehat{\sigma}_\mu^2}\right]}.$$

After plugging in the expressions for both $\widehat{\sigma}_\mu^2$ and $\widehat{\sigma}^2$ above, we obtain

$$2\log\Lambda(\mu; D) = n\log\left[\frac{\widehat{\sigma}_\mu^2}{\widehat{\sigma}^2}\right] = n\log\left[\frac{\sum(X_i - \mu)^2}{\sum(X_i - \bar{X})^2}\right]. \tag{7.14}$$

Compare this with Equation (7.11). The likelihood ratio changes substantially whether σ^2 is known or not. □

Remark 7.3. The parameter ψ is called a *nuisance parameter*. We are interested only in θ, but, in order to study it, we must "take care" of this other unknown parameter in our procedure as well; hence the name, "nuisance". The existence of nuisance parameters often has significant impact on statistical procedures. If the number of nuisance parameters is very large, especially relative to the number of parameters we are actually interested in, "standard" statistical procedures can misbehave very badly. For instance, the MLE of θ can become heavily biased; or the confidence interval for θ can have a much lower probability of containing θ than its nominal level; and so on. They can be a big nuisance indeed. □

The following theorem, due to Samuel S. Wilks, now fully reveals the "magical power" of the twice-logarithmic-transform.

Theorem 2 (Wilks's theorem). *Let $\Lambda_n(\theta; D)$ denote the likelihood ratio given by Equation (7.13), where θ is the true (but unknown) parameter of interest.*

Then, under some regularity conditions and as n → ∞, the distribution of $2 \log \Lambda_n(\theta; D)$ *converges to that of* $\chi^2_{(df)}$, *where* $df = \dim(\theta)$. □

A few clarifications are in order. First, the definition of the $\chi^2_{(df)}$ distribution is given in Exercise 4.4 (section 4.3.2). Next, since the statement of Theorem 2 is asymptotic as $n \to \infty$, the notation "Λ_n" is used (rather than just "Λ") to emphasize the fact that it is based on a total of n i.i.d. random variables. Finally, the regularity conditions consist of various technical assumptions having to do with the differentiability of the likelihood function and the true parameter being an interior point of the parameter space, among others—details which we will (rightfully) skip in this book.

Theorem 2 provides the critical piece to make the general recipe based on $2 \log \Lambda(\theta; D)$,

$$\left\{ \theta : \ 2[\ell(\widehat{\theta}_{mle}, \widehat{\psi}_{mle}) - \ell(\theta, \widehat{\psi}_{mle|\theta})] < c_\alpha \right\}, \tag{7.15}$$

practical. One simply finds c_α by solving $\mathbb{P}(\chi^2_{(df)} < c_\alpha) = 1 - \alpha$; see Figure 7.2.

Remark 7.4. While a great variety of problems satisfy the regularity conditions of Theorem 2, making Equation (7.15) a widely applicable approach, still it is not truly "generic" because circumstances certainly exist when these technical conditions are violated. The finite-mixture model mentioned in section 5.3.2 and Appendix 5.A is one such example. □

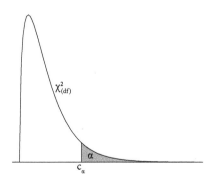

		c_α		
			α	
df	0.100	0.050	0.025	0.010
1	2.71	3.84	5.02	6.64
2	4.61	5.99	7.38	9.21
3	6.25	7.82	9.35	11.35
4	7.78	9.49	11.14	13.28
5	9.24	11.07	12.83	15.09
6	10.65	12.59	14.45	16.81
7	12.02	14.07	16.01	18.48
8	13.36	15.51	17.53	20.09
9	14.68	16.92	19.02	21.67
10	15.99	18.31	20.48	23.21

Figure 7.2 Threshold c_α for the $\chi^2_{(df)}$ distribution such that $\mathbb{P}(\chi^2_{(df)} < c_\alpha) = 1 - \alpha$.

Source: authors.

Example 7.6. Continue with Example 7.5. Equation (7.14) is still a rather daunting function of $X_1, X_2, \ldots, X_n$. Imagine trying to find its distribution, without which we won't be able to determine the proper threshold c_α in (7.15). In situations like this, we really begin to appreciate the value of Theorem 2. We can simply declare that the quantity given in Equation (7.14) is approximately distributed as $\chi^2_{(1)}$—provided that n is not too small. Since $\mathbb{P}(\chi^2_{(1)} < 3.84) = 0.95$ (see Figure 7.2), a 95% confidence interval for μ based on the likelihood ratio is, therefore,

$$\left\{ \mu : n\log\left[\frac{\sum(X_i - \mu)^2}{\sum(X_i - \bar{X})^2} \right] < 3.84 \right\}. \tag{7.16}$$

Notice that our likelihood-ratio strategy has given a very different answer from the one based on the t-distribution, for example, Equation (7.6); the relationship between these two different answers (to the same question) will be further investigated in Exercise 7.2 below. □

Exercise 7.2. This exercise continues with Examples 7.3, 7.5, and 7.6.

(a) First, prove the identity that

$$\sum_{i=1}^{n}(X_i - \mu)^2 = \sum_{i=1}^{n}(X_i - \bar{X})^2 + n(\bar{X} - \mu)^2.$$

(b) Let $T(\mu)$ be defined as Equation (7.5). Starting from Equation (7.14), now show that

$$2\log\Lambda(\mu) = n\log\left[1 + \frac{(T(\mu))^2}{n-1} \right] \longrightarrow (T(\mu))^2 \text{ as } n \to \infty.$$

[*Think: What does this say about the two different confidence intervals— namely, $\{\mu : 2\log\Lambda(\mu) < c_\alpha\}$ and $\{\mu : c_\alpha^{lo} < T(\mu) < c_\alpha^{hi}\}$—for the same underlying problem? Convince yourself by computing both (7.6) and (7.16) on any common set of numbers, $\{X_1, \ldots, X_n\}$ where n is not "too small", and comparing the results.*] □

Exercise 7.3. Recall Example 5.5 (section 5.3). Use the likelihood ratio to find a 95% confidence interval for θ based on data in Table 5.1 (section 5.3). □

7.3.2 Bootstrap

By now, it should be clear to the reader that the main difficulty in coming up with a usable pivotal quantity $h(\theta; D)$ is the second requirement (see the beginning of section 7.3); the first requirement is relatively trivial. Can we "overcome" this difficulty with a numeric trick of some sort? The answer is: yes, we can do this by using the *bootstrap* method.

If we repeatedly sample *with replacement* from the data D, we can get many slightly different data samples $D^{*(1)}, D^{*(2)}, \ldots, D^{*(B)}$ of the same size n. The main idea of the bootstrap is that we can approximate the distribution of $h(\theta; D)$ with the empirical distribution[1] of

$$\left\{ h(\widehat{\theta}; D^{*(1)}), h(\widehat{\theta}; D^{*(2)}), \ldots, h(\widehat{\theta}; D^{*(B)}) \right\}.$$

Exercise 7.5 below will allow readers to verify this in a setting which should be familiar by now. From this empirical approximation, we can identify a set C_α^*—e.g., $(c_\alpha^{*lo}, c_\alpha^{*hi})$—such that

$$\frac{1}{B} \sum_{b=1}^{B} I\left\{ h(\widehat{\theta}; D^{*(b)}) \in C_\alpha^* \right\} = 1 - \alpha,$$

where $I(\cdot)$ is a binary indicator function. In the one-dimensional case, this can be done by relying on the empirical quantiles of the aforementioned histogram. In multidimensional cases, this can be more complicated but the underlying principle is the same; see Figure 7.3 for an illustration.

Afterwards, we simply declare that

$$\mathbb{P}(h(\theta; D) \in C_\alpha^*) \approx 1 - \alpha,$$

so that

$$h^{-1}(C_\alpha^*; D) \equiv \{\theta : h(\theta; D) \in C_\alpha^*\}$$

is the desired answer.

Example 7.7. As we are no longer constrained by the need to work out its distribution theoretically, certainly nothing will prevent us from using what may be arguably the simplest function $h(\theta; D) = \widehat{\theta} - \theta$, as we did in Example 7.2. [*Note: Recall that $\widehat{\theta}$ is a function of the data D alone.*]

---

[1] Think of the histogram.

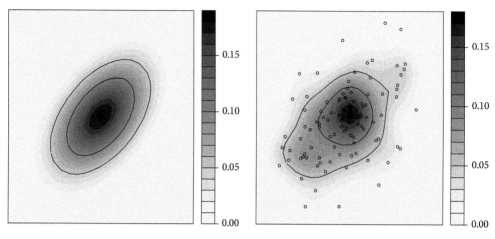

Figure 7.3   Finding the sets, C_α on the left and C_α^* on the right, such that $\mathbb{P}(h(\theta; \boldsymbol{D}) \in C_\alpha) = (1/B) \sum_{b=1}^{B} I(h(\widehat{\theta}; \boldsymbol{D}^{*(b)}) \in C_\alpha^*) = 1 - \alpha$, for two different levels of α.

Note: The shaded scale at any given location is proportional to the density of the distribution (left) or the number of points in its immediate neighborhood (right).

Source: authors.

This means

$$h(\widehat{\theta}; \boldsymbol{D}^{*(b)}) = \widehat{\theta}^{*(b)} - \widehat{\theta},$$

where $\widehat{\theta}^{*(b)}$ denotes the estimate of θ obtained by applying the same procedure (as the one which produced $\widehat{\theta}$ on the original data $\boldsymbol{D}$) to the resampled data $\boldsymbol{D}^{*(b)}$. [*Think: This statement can be confusing for beginners. Make sure you truly understand what it means.*]

Next, we find $c_\alpha^{*\text{lo}}, c_\alpha^{*\text{hi}}$ to be the empirical quantiles of $\{\widehat{\theta}^{*(b)} - \widehat{\theta}\}_{b=1}^{B}$. As a concrete illustration, suppose $\alpha = 0.05$ and $B = 1000$. In principle, this task is now easily achievable by letting

$$c_{0.05}^{*\text{lo}} = \text{the 25-th } \textit{smallest} \quad \text{and} \quad c_{0.05}^{*\text{hi}} = \text{the 25-th } \textit{largest}$$

number among $\{\widehat{\theta}^{*(1)} - \widehat{\theta}, \ldots, \widehat{\theta}^{*(1000)} - \widehat{\theta}\}$. [*Note: In practice, there are different interpolating and tie-breaking algorithms for finding empirical quantiles; they are useful especially if B is chosen to be relatively small, which may be necessary if the estimation procedure for producing $\widehat{\theta}$ from $\boldsymbol{D}$, or $\widehat{\theta}^{*(b)}$ from $\boldsymbol{D}^{*(b)}$, is computationally expensive.*] This leads to the confidence interval:

$$\{\theta : c_\alpha^{*\text{lo}} < \widehat{\theta} - \theta < c_\alpha^{*\text{hi}}\} \Leftrightarrow (\widehat{\theta} - c_\alpha^{*\text{hi}}, \widehat{\theta} - c_\alpha^{*\text{lo}}). \tag{7.17}$$

Let q_α^{*lo}, q_α^{*hi} denote the empirical quantiles of $\{\widehat{\theta}^{*(b)}\}_{b=1}^B$. Clearly,

$$c_\alpha^{*lo} = q_\alpha^{*lo} - \widehat{\theta} \quad \text{and} \quad c_\alpha^{*hi} = q_\alpha^{*hi} - \widehat{\theta}$$

as the two samples merely differ from each other by a common shift (of $\widehat{\theta}$). This means the confidence interval (7.17) can be equivalently expressed as

$$(\widehat{\theta} - c_\alpha^{*hi}, \; \widehat{\theta} - c_\alpha^{*lo}) \Leftrightarrow (2\widehat{\theta} - q_\alpha^{*hi}, \; 2\widehat{\theta} - q_\alpha^{*lo}). \tag{7.18}$$

This can be a rather confusing result. As a matter of fact, many people who use the bootstrap to construct confidence intervals do not really understand where the "additional" factor of 2 comes from. Beware of which quantiles are being used! □

Exercise 7.4. Recall, once again, Example 5.5 in section 5.3. Use the bootstrap, by following Example 7.7, to find a 95% confidence interval for θ based on data in Table 5.1 (section 5.3), and compare with the one you obtained earlier in Exercise 7.3. □

Exercise 7.5. Let

$$h(\theta; D) = 2[\ell(\widehat{\theta}; D) - \ell(\theta; D)]$$

be the twice-log-transformed likelihood ratio, where $\widehat{\theta}$ is the MLE of θ based on D. Verify empirically that the bootstrap and Wilks's theorem (Theorem 2) are in agreement about how this quantity is distributed by using (again!) the likelihood from Example 5.5 and data in Table 5.1 (section 5.3). [*Hint: Let* $\widehat{\theta}^{*(b)}$ *denote the MLE of* θ *based on the resampled data* $D^{*(b)}$. *Which of the following,*

$$\begin{aligned}
h(\widehat{\theta}; D^{*(b)}) &= 2[\ell(\widehat{\theta}^{*(b)}; D) - \ell(\widehat{\theta}; D)], \\
h(\widehat{\theta}; D^{*(b)}) &= 2[\ell(\widehat{\theta}^{*(b)}; D^{*(b)}) - \ell(\widehat{\theta}; D)], \\
h(\widehat{\theta}; D^{*(b)}) &= 2[\ell(\widehat{\theta}^{*(b)}; D^{*(b)}) - \ell(\widehat{\theta}; D^{*(b)})],
\end{aligned}$$

is the correct way to implement the bootstrap?] □

Of course, the bootstrap can be used to do *a lot* more than just coming up with confidence intervals. For example, the bias of an estimator $\widehat{\theta}$ usually cannot be computed directly since the true value of the parameter θ is unknown, but by choosing $h(\theta; D) = \widehat{\theta} - \theta$, we can even use the bootstrap to estimate this otherwise unknowable quantity by

$$\mathbb{E}(\widehat{\theta}) - \theta = \mathbb{E}(\underbrace{\widehat{\theta} - \theta}_{h(\theta;\, D)}) \approx \frac{1}{B} \sum_{b=1}^{B} (\underbrace{\widehat{\theta}^{*(b)} - \widehat{\theta}}_{h(\widehat{\theta};\, D^{*(b)})}) = \left[\frac{1}{B} \sum_{b=1}^{B} \widehat{\theta}^{*(b)} \right] - \widehat{\theta}.$$

An important "special case" corresponds to choosing $h(\theta; D) = h(D) = h(X_1, \ldots, X_n)$ to be "merely" a function of the data set alone. Recall from section 3.2.1 that the distribution of $h(X_1, \ldots, X_n)$ is generally hard to derive, but the bootstrap still gives us a way to estimate it. For example, since any estimator $\widehat{\theta}$ of θ is a function of the data set alone, we can always use the bootstrap to estimate its variance with

$$\mathbb{V}\mathrm{ar}(\widehat{\theta}) \approx \frac{1}{B-1} \sum_{b=1}^{B} (\widehat{\theta}^{*(b)} - \overline{\theta^*})^2, \quad \text{where} \quad \overline{\theta^*} \equiv \frac{1}{B} \sum_{b=1}^{B} \widehat{\theta}^{*(b)}.$$

We can even choose $h(D)$ to be a predictive quantity such as the one given by Equation (7.1), and the bootstrap readily allows us to gauge the amount of uncertainty in our prediction—by examining the empirical distribution of

$$h(D^{*(b)}) = \int_{x \in A} f(x; \widehat{\theta}^{*(b)}) dx, \quad b = 1, 2, \ldots, B.$$

Remark 7.5. In principle, it may also appear as if the bootstrap "should just work" for any choice of $h(\theta; D)$. In practice, however, not all choices of $h(\theta; D)$ will work equally well; some will work better than others. For example, to find confidence intervals, the choice $h(\theta, D) = \widehat{\theta} - \theta$, which we used in Example 7.7, is certainly the most straightforward but not necessarily the best. (Unfortunately, we won't be able to go into these nuances in this book.) A similar illusion exists for the Metropolis algorithm (see Remark 6.5 in section 6.A.2)—while it may appear to work in principle for "any" proposal distribution, in practice not all proposal distributions are the same; see Exercise 6.5 (section 6.A.2). $\qquad\square$

8
Tests of Significance

Another, seemingly different, class of inferential problems, which also requires us to take into explicit consideration the amount of uncertainty we have about θ, is known as *tests of significance*. The main objective is to decide whether a particular statement about the model parameter θ—usually referred to as the *null hypothesis* and denoted as H_0—is supported by data. Although the null hypothesis can take on many different forms (see, e.g. Remark 8.2 and Exercise 8.2 later on), by far the most common (and also the simplest) one is $H_0 : \theta = \theta_0$.

Clearly, such inferential problems only make sense when the parameter itself has some special meaning. For instance, in Example 7.1 at the beginning of Chapter 7, scientists may be interested in the special parameter value of zero (i.e. a null hypothesis of $H_0 : \beta = 0$), which corresponds to the proposition that sleep appears to have no effect on obesity.

This illustrates, to a certain extent, why tests of significance have seen wide applications in almost all disciplines of science—because they provide a paradigm for answering important empirical questions such as, "Does sleep appear to affect obesity?", "Is drinking coffee regularly associated with a lower risk of dementia?", "Does a university degree have any apparent impact on lifetime earnings?" and so on.

But it also explains why we do not come across significance testing in fields like machine learning as much as we do in other scientific disciplines. When using a deep neural network with millions of parameters to decide whether an image contains a cat or not, for example, we are not usually interested in whether a specific parameter is equal to any special value. These parameters do not have any intrinsic meaning in themselves. As such, there is no special value for us to pay attention to—such as $\beta = 0$ in Example 7.1—and we just want to estimate all the parameters well enough so the model can be used to make the necessary predictions as accurately as possible.

Remark 8.1. Later in section 8.2, we will explain that this paradigm itself is currently facing some unprecedented challenges. Nonetheless, it still remains, for historic reasons, the dominant paradigm for conducting

Essential Statistics for Data Science. Mu Zhu, Oxford University Press. © Mu Zhu (2023).
DOI: 10.1093/oso/9780192867735.003.0008

empirical investigations. That's why it is impossible to study statistics without paying at least some attention to this topic. □

8.1 Basics

An intuitive way to proceed would be to first estimate the unknown parameter θ and then check whether the resulting estimate $\hat{\theta}$ is close enough to θ_0. But how close is close enough?

Broadly speaking, we would like to construct a quantity, $t(\theta; D)$, called a *test statistic* in this context, so that the test decision can be made according to the following rule:

$$\text{reject } H_0 \quad \text{if} \quad t(\theta_0; D) \geq c, \tag{8.1}$$

where c is a decision threshold.

The intuitive approach of checking whether $\hat{\theta}$ is close to the hypothesized value θ_0 would amount to using $t(\theta_0; D) = |\hat{\theta} - \theta_0|$ as the test statistic, and the question of how close is close enough would amount to how the decision threshold c should be chosen. This is where uncertainty begins to matter. If we are very certain in our estimation, we may think even $|\hat{\theta} - \theta_0| = 0.01$ is still not close enough, whereas if we are very uncertain in our estimation, we may think even $|\hat{\theta} - \theta_0| = 100$ is already close enough.

Definition 8 (Significance level). *The probability of a test incorrectly rejecting H_0 when H_0 is actually true is called its significance level.* □

By convention, one chooses the decision threshold c in a test so as to control its significance level at, say, α. (Typical choices are $\alpha = 0.05$ or $\alpha = 0.01$.) The corresponding threshold is often denoted "c_α" to emphasize its dependence on the desired significance level α, and it satisfies the equation,

$$\mathbb{P}[t(\theta_0; D) \geq c_\alpha] = \alpha. \tag{8.2}$$

As the name suggests, the null hypothesis is often a statement that says "nothing special" is happening—it is a statement of the status quo. Hence, all tests of significance are based upon the philosophical premise that the status quo should always be accepted unless there is strong evidence to suggest otherwise.[1] In other words, one must be presumed innocent until proven guilty beyond a reasonable doubt.

8.1.1 Relation to interval estimation

Like the pivotal quantity $h(\theta; D)$ we encountered earlier (section 7.3), to make decisions with the test statistic $t(\theta; D)$ we must also be able to (i) evaluate it given the parameter and the data alone—so that we can assess the hypothesized parameter value agaist the data, and (ii) know its distribution—so that we can determine the decision threshold by solving (8.2).

In fact, if a certain pivotal quantity allows us to construct confidence intervals, we can use it to conduct significant tests (about the same parameter); and if a certain test statistic allows us to conduct significance tests, we can use it to construct confidence intervals (again, for the same parameter). A direct comparison of Equations (7.7) and (8.2) reveals that these two inferential problems are actually complementary to each other in the sense made more precise by Proposition 1 below. (That's why we started this section by calling significance tests a "seemingly different" class of inferential problems.)

Proposition 1. *If a test, which rejects $H_0 : \theta = \theta_0$ when $t(\theta_0; D) \geq c_\alpha$, has a significance level of α, then $\{\theta : t(\theta; D) < c_\alpha\}$ will contain the true parameter with probability $(1 - \alpha) \times 100\%$. Conversely, if $\{\theta : h(\theta; D) \in C_\alpha\}$ contains the true parameter with probability $(1 - \alpha) \times 100\%$, then a test, which rejects $H_0 : \theta = \theta_0$ when $h(\theta_0; D) \notin C_\alpha$, will have a significance level of α.*

Exercise 8.1. Prove Proposition 1. □

Example 8.1. Recall Example 7.3 (section 7.2). The test that rejects $H_0 : \mu = \mu_0$ if μ_0 falls outside the confidence interval, that is, if

$$\frac{|\bar{X} - \mu_0|}{S/\sqrt{n}} \geq c_\alpha,$$

is called a (one-sample) *t-test*.[2] □

---

[1] In scientific investigations, the status quo is often the latest scientific theory, or our current understanding of the state of nature, and the purpose of a significance test is to decide whether there is evidence to challenge our existing knowledge.

[2] The *t*-test has many variations. In model (5.9) in section 5.1.2, for example, we can test $H_0 : \beta = 0$ with a (slightly different) *t*-test, too.

Example 8.2. Another widely used test is the *likelihood ratio test* (LRT), which rejects $H_0 : \theta = \theta_0$ if θ_0 falls outside the region defined by Equation (7.15), that is, if

$$2[\ell(\widehat{\theta}_{mle}, \widehat{\psi}_{mle}) - \ell(\theta_0, \widehat{\psi}_{mle|\theta_0})] \geq c_\alpha.$$

Clearly, a similar statement applies to the simpler Equation (7.12) as well, when there is no nuisance parameter. □

Remark 8.2. Quite often in practice, the null hypothesis H_0 does not necessarily specify explicitly what the parameter is equal to; instead, it merely imposes additional constraints on the parameter space, that is, $H_0 : \theta \in \Omega_0$. (The explicit case corresponds to the set $\Omega_0 = \{\theta_0\}$ containing just one point.)

A fairly common example of this type for $\theta \equiv (\theta_1, \theta_2)^\top \in \mathbb{R}^2$ is $\Omega_0 = \{(\theta_1, \theta_2) : \theta_1 = \theta_2\}$. In other words, the null hypothesis states that two different parameters θ_1 and θ_2 are equal (see Exercise 8.2 below), which translates into a restriction that the corresponding two-dimensional parameter θ can only lie on the 45-degree line instead of in the entire space of $\mathbb{R}^2$.

The LRT has the advantage of being easily adaptable to such problems, and Wilks's theorem still holds. Instead of evaluating Equation (7.13) "at a specific $\theta = \theta_0$", which we can no longer do, we simply evaluate it "on the set Ω_0", meaning

$$\Lambda \equiv \frac{\sup\limits_{\theta,\psi\in\Omega_0^c} L(\theta,\psi)}{\sup\limits_{\theta,\psi\in\Omega_0} L(\theta,\psi)} = \frac{L(\widehat{\theta}_{mle|\Omega_0^c}, \widehat{\psi}_{mle|\Omega_0^c})}{L(\widehat{\theta}_{mle|\Omega_0}, \widehat{\psi}_{mle|\Omega_0})},$$

where $\widehat{\theta}_{mle|A}$ (or $\widehat{\psi}_{mle|A}$) denotes the constrained MLE of θ (or of ψ) in the restricted set A, and compute the degree of freedom as

$$df = \dim(\Omega_0^c) - \dim(\Omega_0).$$

For example, suppose there is no nuisance parameter ψ, $\theta = (\theta_1, \theta_2)^\top \in \mathbb{R}^2$, and the null hypothesis is $H_0 : \theta_1 = \theta_2$. Then, $\dim(\Omega_0^c) = 2$ since, outside of Ω_0, there are two separate parameters; and $\dim(\Omega_0) = 1$ since, when restricted to Ω_0, there is just one (common) parameter. Generally, for $H_0 : \theta \in \Omega_0 = \{\theta_0\}$, we would simply have $\dim(\Omega_0) = \dim(\psi)$, since the parameter θ has been completely specified (i.e. fixed) by Ω_0, or $\dim(\Omega_0) = 0$ if there is no nuisance parameter ψ. ☐

Exercise 8.2. Suppose there are two sets of i.i.d. random variables,

$$\mathcal{X} = \left\{X_1,\ldots,X_n \overset{iid}{\sim} \text{Poisson}(\lambda_1)\right\} \quad\text{and}\quad \mathcal{Y} = \left\{Y_1,\ldots,Y_m \overset{iid}{\sim} \text{Poisson}(\lambda_2)\right\}.$$

Furthermore, the two sets $\mathcal{X}$ and $\mathcal{Y}$ are independent of each other as well. For example, each X_i may be the number of seizures reported by a patient taking a new drug and each Y_j, the same number reported by a patient taking a placebo. We are interested in testing the null hypothesis

$$H_0: \quad \lambda_1 = \lambda_2$$

or whether there is any difference between the new drug and the placebo.

(a) Show that the likelihood ratio test will reject H_0 if

$$2\left[(n\bar{X})\log\frac{(n+m)\bar{X}}{n\bar{X}+m\bar{Y}} + (m\bar{Y})\log\frac{(n+m)\bar{Y}}{n\bar{X}+m\bar{Y}}\right] \geq c$$

for some threshold c, and determine the value of c so that the test has a significance level of approximately 0.01.
(b) How would you do this using the bootstrap instead?

[*Hint: For (b), let $\theta \equiv \lambda_1 - \lambda_2$. What does H_0 say about θ, and how can you construct a confidence interval for it?*] ☐

8.1.2 The p-value

If one counts the occurrence of statistical jargon in the literature of all experimental and observational sciences (whether it is astronomy, biology, chemistry, medicine, psychology, sociology, or zoology), the top-ranking jargon will probably be the p-value.

Definition 9 (p-value). *The p-value is defined as*

$$p\text{-value} \equiv \mathbb{P}[t(\theta_0; D) \geq t_{obs}],$$

where $t_{obs} \equiv t(\theta_0; D_{obs})$ *denotes the numeric value of the test statistic* $t(\theta; D)$ *evaluated at the null hypothesis* H_0 *and the observed data.*   □

Figure 8.1 shows the concept graphically. Since, by Equation (8.2), the decision threshold c_α is chosen so that $\mathbb{P}[t(\theta_0; D) \geq c_\alpha] = \alpha$, it is not hard to infer from both Definition 9 and Figure 8.1 that

$$t_{obs} < c_\alpha \;\Rightarrow\; \text{p-value} > \alpha;$$

$$t_{obs} = c_\alpha \;\Rightarrow\; \text{p-value} = \alpha;$$

$$t_{obs} > c_\alpha \;\Rightarrow\; \text{p-value} < \alpha.$$

Therefore, at a significance level of α, the decision rule (8.1) is the same as

$$\text{reject } H_0 \quad \text{if} \quad \text{p-value} \leq \alpha. \tag{8.3}$$

The p-value is commonly interpreted as the probability of seeing a measurement (the test statistic) to be this extreme ($\geq t_{obs}$) under the status quo (H_0). If such a measurement is supposed to be fairly common (large p-value), then there is nothing unusual about seeing one; but if it is supposed to be quite rare (small p-value), then seeing one naturally raises our suspicion that there may be something wrong with the status quo.

The main attraction of the p-value is that, instead of a simple binary decision of whether the null hypothesis is rejected or not, it provides a universal numeric scale—regardless of the choice of the test statistic—to quantify the strength of statistical evidence against the null. The smaller the p-value, the stronger the evidence. (See also Exercise 8.3 below.)

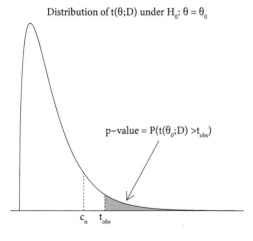

Distribution of $t(\theta;D)$ under H_0: $\theta = \theta_0$

p–value = $P(t(\theta_0;D) > t_{obs})$

c_α t_{obs}

Figure 8.1 The p-value measures, under the null hypothesis, the probability of the test statistic being as extreme as it is on the observed data set.

Source: authors.

That's why almost all scholarly journals require scientists to report p-values for their experiments. In fact, many journals will refuse to publish a study if its p-value is larger than the conventional cut-off value of 0.05—presumably because the statistical evidence is not strong enough to justify a challenge of the status quo (H_0); in other words, the statistical evidence is not strong enough that the study has found anything new or interesting. Unfortunately, such practice has led to much abuse.

Exercise 8.3.   Let $F_0(t)$ denote the cumulative distribution function of a test statistic, T, under the assumption that the null hypothesis H_0 is true. Define a new random variable $U = 1 - F_0(T)$. What distribution does U follow under H_0? [*Think: By definition, the p-value is the probability* $\mathbb{P}(T \geq t_{obs}) = 1 - F_0(t_{obs})$ *under* H_0. *What can you learn from this exercise about the p-value? For example, if we repeated the same hypothesis test with different data (from the same data-generating process), each time we would obtain a slightly different p-value. How would these p-values be distributed under* H_0? *What fraction of them would fall below 0.05, between 0.2 and 0.3, and above 0.9, respectively?*]   □

8.2 Some challenges

As we briefly mentioned earlier (Remark 8.1 at the beginning of the chapter), the paradigm of significance testing is facing some big challenges. In this section, we briefly discuss some of them.

8.2.1 Multiple testing

The first challenge is encapsulated by the following simple calculation. Suppose we conduct N independent significance tests, each at a significance level of $\alpha = 0.05$. If all the null hypotheses are true, then

$\mathbb{P}(at\ least\ one\ \text{false rejection})$

$$= 1 - \mathbb{P}(no\ \text{false rejection}) = 1 - (1 - \alpha)^N$$

$$\underset{\underset{\alpha = 0.05}{\uparrow}}{=} \begin{cases} 0.226, & \text{if } N = 5; \\ 0.401, & \text{if } N = 10; \\ \vdots & \vdots \\ 0.923, & \text{if } N = 50. \end{cases} \tag{8.4}$$

Consider again the context of Example 7.1. A curious scientist may be interested not only in the effect of sleep on obesity but also in the effect of many other factors, for example, alcohol, children, friends, marriage, smoking, climate and so on. Suppose this curious scientist is well funded and conducts a large number of independent studies[3] to test the effect of each of these different factors on obesity. Then, even if none of these factors has any real effect on obesity, the calculations contained in Equation (8.4) imply that there is still a very high chance that this scientist will end up rejecting at least one of his/her null hypotheses and concluding erroneously that a certain factor which he/she has investigated—such as friends—may appear to have some effect on obesity.

This is referred to as the *multiple testing problem* in statistics, and it explains, to a certain extent, why some scientific disciplines are currently experiencing a so-called "replication crisis".

These days, we are bombarded by headlines such as "Carrots have hidden healing powers", "Coffee is good for you", or, wait—"Coffee is bad for you". Huh? Sounds familiar? An article [14] in the *Significance* magazine reported that, between 1990 and 2010, 12 clinical trials were conducted to test 52 scientific claims about the health benefits (or hazards) of vitamin E, vitamin D, calcium, selenium, hormone replacement therapy, folic acid, beta-carotene and so on. And guess what? Not a single one of them could be replicated!

When those 52 claims were first published, all of the original studies reported a p-value of less than 0.05. This was *supposed to* give us the assurance that each of those claims had less than 5% chance of turning out to be a false discovery. Assuming that these studies were independent, we should expect only about $52 \times 5\% \approx 3$ mistakes of this kind and not replicable as a result. In fact,

$$\mathbb{P}[\text{Binomial}(52, 0.05) > 20] \approx 2.0 \times 10^{-14},$$

so having more than 20 of them failing to be replicated should have been almost impossible. The fact that *all* of them failed to replicate suggests strongly that the original studies had much higher chances of being false discoveries on average than the nominally reported 5% or less.

The multiple testing problem is especially serious in our age of big data. For instance, we are now routinely testing the entire human genome, which is believed to contain about 20,000 protein-encoding genes when this chapter is

---

[3] In reality, the chances are that these studies won't be entirely independent of each other, but we will ignore this complication here.

being written, against various phenotypes, such as being obese. That's 20,000 significance tests altogether, not just 10 or 50. Many genes will be found to have a statistical association with the phenotype—that is, the corresponding null hypothesis that the underlying gene has no effect will be rejected, but many of these rejections will be false discoveries and merely spurious genetic associations.

That's why scholarly journals now use a much lower p-value threshold for genomic studies. One conservative correction, called the *Bonferroni correction*, requires that each individual test use a significance level of α/N when a total of N significance tests are conducted. This ensures

$$\mathbb{P}(\textit{at least one} \text{ false rejection}) \leq \sum_{j=1}^{N} \mathbb{P}(\text{false rejection by the } j\text{-th test})$$

$$= N \times (\alpha/N)$$

$$= \alpha,$$

where the first inequality is based on a result given in Exercise 8.4 below. For a less conservative procedure, see Fun Box 4 below.

Fun Box 4

The Benjamini-Hockberg testing procedure. For $i = 1, 2, \ldots, N$, let u_i denote the p-value associated with testing the i-th null hypothesis, $H_{0,i}$. Order everything in decreasing significance (increasing p-value), as shown below:

$$H_{0,(1)} \; > \; H_{0,(2)} \; > \; \ldots \; > \; H_{0,(N)}$$
$$u_{(1)} \; < \; u_{(2)} \; < \; \ldots \; < \; u_{(N)}.$$

Let

$$i_{max} \equiv \max_{1, \ldots, N} \{i : u_{(i)} \leq i\alpha/N\}$$

be the largest index for which the inequality above holds. Reject $H_{0,(i)}$ for all $i \leq i_{max}$; accept the rest. [*Think: Why is this less conservative than the Bonferroni correction?*] An intuitive understanding of this procedure actually requires a fairly fundamental conceptual shift in how we think about significance testing; see Appendix 8.A.

Lowering the p-value cut-off is possible for genomic studies because the size of the human genome—and hence the total number of significance tests a scientist may conduct—is presumably the same for every scientist, which allows the community to agree on a commonly applicable correction, but such universal adjustments are not always practical. Conducting many significance tests in secret and publishing only those with p-values less than the required cut-off (e.g. 0.05) continues to be a serious abuse that is not easy to curb or eradicate. We are in dire need of a good solution.[4]

Exercise 8.4. Draw a Venn diagram to convince yourself that

$$\mathbb{P}\left(\bigcup_{j=1}^{N} A_j\right) \le \sum_{j=1}^{N} \mathbb{P}(A_j).$$

[*Note: This is known as Boole's inequality.*]   □

8.2.2 Six degrees of separation

Another challenge arises from the widely shared experience among many statisticians and data analysts that most null hypotheses end up being rejected when we have a lot of data, even when it is fairly clear that they shouldn't be. In this section, we describe a simple thought experiment to offer a conjecture of why this may be so.

Suppose a group of people all have some knowledge about a certain value, μ, and our plan is to ask everyone to report what he/she knows—say $X_1, X_2, \ldots, X_n$—and apply Equation (7.6) to obtain a confidence interval for μ:

$$\left(\bar{X} - 2\sqrt{\frac{S^2}{n}}, \bar{X} + 2\sqrt{\frac{S^2}{n}}\right), \quad \text{where} \quad S^2 = \frac{1}{n-1}\sum_{i=1}^{n}(X_i - \bar{X})^2.$$

As the width of this interval is random (see also Remark 7.2; section 7.2), we will simplify our analysis by considering a hypothetical interval of the "average width":

$$C_\mu \equiv \left(\bar{X} - 2\sqrt{\mathbb{E}\left[\frac{S^2}{n}\right]}, \bar{X} + 2\sqrt{\mathbb{E}\left[\frac{S^2}{n}\right]}\right).$$

---

[4] At the level of the current book, it is impossible to go much deeper. While it is true that we don't yet have genuinely satisfactory or widely acceptable solutions to this problem, it is also important for us to state that many research efforts are being made to address it, and what we have touched upon in this book is merely the tip of the iceberg.

If $X_1, X_2, \ldots, X_n \overset{iid}{\sim} N(\mu, \sigma^2)$, then $\mathbb{E}(S^2) = \sigma^2$ (see Exercise 5.7; section 5.2) and the probability of C_μ missing—or that of the corresponding significance test falsely rejecting—the true value of μ should be about 5%. But the story can be very different if these random variables are *not* independent.

For instance, let us consider a highly simplified network model for spreading information (Figure 8.2), which we will denote by $\mathcal{G}(L; \mu, \sigma^2)$. Here is how it works. One person, who knows the value of μ perfectly, communicates this information to two friends; each friend, in turn, communicates what he or she knows to two other friends, and the process continues recursively. In other words, the value μ is spread like a virus in a network, and the topology of the network is that of a simple binary tree. Here is the main catch: each time the value is communicated, the information is contaminated by an independent random noise from $N(0, \sigma^2)$. The parameter L refers to the number of layers in the network.

Under such a model, it can be shown (see Exercise 8.5)[5] that the probability of C_μ missing—or that of the corresponding significance test falsely rejecting—the true value of μ will converge upwards to 100% as $L \to \infty$ (that is, the more "information" we gather, the worse-off we are) and we will practically always end up rejecting the true value.

As it has often been said, we live in a small world and there are only about six degrees of separation between any two of us. What we have shown above

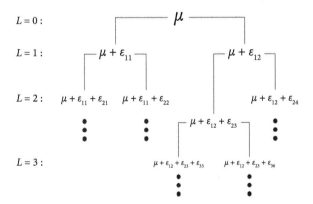

Figure 8.2 The network, $\mathcal{G}(L; \mu, \sigma^2)$, in which the value μ is communicated with error.

Note: All $\varepsilon_{\ell j}$, for $\ell = 1, 2, \ldots, L$ and $j = 1, 2, 3, \ldots, 2^\ell$, are independent random noises from $N(0, \sigma^2)$.

Source: authors.

---

[5] This exercise is slightly beyond the scope of this book.

is an example of how the assumption that $X_1, X_2, \ldots, X_n$ are independent can really throw us off the track on large data sets. Yes, we do appear to be rejecting almost any null hypothesis when we have a lot of data, but what we are really rejecting is our assumption that the random variables generated by the probability model are independent of each other. As our example has illustrated, with lots of data, we may simply have mounting evidence that such an assumption has little chance of being correct in the small world in which we live.

Remark 8.3.   How to specify the joint distribution of $X_1, X_2, \ldots, X_n$ without assuming they are independent of each other is not at all an easy task. Usually, this would be possible only if we had some additional information about them. For example, we may know that these random variables are generated in a sequential order, for example:

$$ X_1 \rightarrow X_2 \rightarrow \ldots \rightarrow X_{n-1} \rightarrow X_n \rightarrow \ldots, $$

so that we may say X_n is almost independent of X_1 but very unlikely to be independent of X_{n-1}. Or we may know that these random variables are spatially organized, for example:

$$
\begin{matrix}
X_{11} & X_{12} & X_{13} & \cdots \\
X_{21} & X_{22} & X_{23} & \cdots \\
X_{31} & X_{32} & X_{33} & \cdots \\
\vdots & \vdots & \vdots & \ddots
\end{matrix}
,
$$

so that we may say X_{22} is dependent on its neighbors $X_{21}, X_{23}, X_{12}, X_{32}$ but, conditional on these neighbors, it is independent of all others. The model $\mathcal{G}(L; \mu, \sigma^2)$, which we considered in this section, is no exception.

What is more, the distribution of $\hat{\theta}$ and that of $\hat{\theta} - \theta$ will be a lot harder to work out as well if the underlying data $D = \{X_1, \ldots, X_n\}$ are not independent. Even the bootstrap (section 7.3.2) will be tricky (though not impossible) to apply. For example, if X_3, X_7, X_{11} are a highly dependent cluster, but X_7, X_{11} are not included in the bootstrap sample $D^{*(b)}$, then the critical dependence structure will be broken, making it difficult for us to draw the correct statistical inference. As is the case with every other research-active discipline, there are always more problems than solutions.   □

Exercise 8.5. Consider the network model $\mathcal{G}(L; \mu, \sigma^2)$ for spreading information. In total, it is clear we have altogether $n_L = 1 + 2 + 4 + \cdots + 2^L$ pieces of
information, which we will collect into a set, say, $D_L \equiv \{X_1, X_2, \ldots, X_{n_L}\}$. Let

$$\bar{X}_L = \frac{1}{n_L} \sum_{X_i \in D_L} X_i \quad \text{and} \quad S_L^2 = \frac{1}{n_L - 1} \sum_{X_i \in D_L} (X_i - \bar{X}_L)^2$$

denote the usual sample mean and sample variance of the collection, D_L.

(a) Show that $\mathbb{E}(\bar{X}_L) = \mu$.

(b) Show, by induction on L, that

$$\mathbb{V}\mathrm{ar}(\bar{X}_L) = \sigma^2 \Omega_L, \quad \text{where} \quad \Omega_L = \frac{4^{L+1} - (2L+1)2^{L+1} - 2}{4^{L+1} - 2 \times 2^{L+1} + 1} \longrightarrow 1$$

$$\text{as } L \to \infty.$$

(c) Show, by induction on L, that

$$\mathbb{E}\left[\frac{S_L^2}{n_L}\right] = \sigma^2 \widetilde{\Omega}_L, \quad \text{where}$$

$$\widetilde{\Omega}_L = \frac{(L-2)4^{L+1} + (L+4)2^{L+1}}{8^{L+1} - 4 \times 4^{L+1} + 5 \times 2^{L+1} - 2} \longrightarrow 0 \quad \text{as } L \to \infty.$$

(d) Let π_L be the probability that the "average width" confidence interval
C_μ—here, constructed from D_L, that is, using $\bar{X}_L$, S_L^2 and n_L—will miss
the true value of μ. Show that

$$\pi_L = 2\left[1 - \Phi\left(2\sqrt{\widetilde{\Omega}_L/\Omega_L}\right)\right] \uparrow 1 \quad \text{as } L \to \infty,$$

where $\Phi(\cdot)$ is the cumulative distribution function of $N(0, 1)$.

[*Hint: In section 3.A.3, we briefly explained why the variance formula for sums
of non-independent random variables is necessarily more complicated and that
it must include all pairwise covariance terms. At the level of this book, we are
primarily dealing with independent random variables, and the very concept of*

*the covariance itself has not really been touched upon, but the following two
results involving the covariance,*

$$\mathbb{V}\mathrm{ar}\left[\sum_{i=1}^{n} a_i X_i\right] = \sum_{i=1}^{n} a_i^2 \mathbb{V}\mathrm{ar}(X_i) + \sum_{i \neq j} a_i a_j \mathbb{C}\mathrm{ov}(X_i, X_j) \qquad (8.5)$$

and

$$\mathbb{C}\mathrm{ov}\left[\sum_{i=1}^{n} a_i X_i, \sum_{j=1}^{m} b_j Y_j\right] = \sum_{i=1}^{n}\sum_{j=1}^{m} a_i b_j \mathbb{C}\mathrm{ov}(X_i, Y_j), \qquad (8.6)$$

*are needed for completing this exercise as the "main catch" of the network
model, $\mathcal{G}(L; \mu, \sigma^2)$, is that it generates non-independent random variables. These
equations are most easily remembered by comparing (8.5) to computing per-
fect squares such as $(a_1 + a_2 + a_3)^2$ and (8.6) to doing multiplications such as
$(a_1 + a_2)(b_1 + b_2)$.]* □

Remark 8.4. As Exercise 8.5 is slightly beyond the scope of this book, we will
describe here a small simulation to "verify" the correctness of all the results
contained in it. For each $L = 1, 2, \ldots, 15$, a total of $B = 500$ simulations were
run. For each $b = 1, 2, \ldots, B$, a set $D_L^{(b)}$ was first generated from $\mathcal{G}(L; 0, 1)$; its
sample mean $\bar{X}_L^{(b)}$ and sample standard deviation $S_L^{(b)}$ were then computed
and recorded. Also recorded was whether the interval

$$I^{(b)} \equiv \left(\bar{X}_L^{(b)} - 2\frac{S_L^{(b)}}{\sqrt{n_L}}, \bar{X}_L^{(b)} + 2\frac{S_L^{(b)}}{\sqrt{n_L}}\right)$$

missed the true value 0. (Notice that here, each $I^{(b)}$ was the "usual" confidence interval rather than the "average width" confidence interval C_μ, which
we used in the theoretical analysis.) Afterwards, the quantity $\mathbb{V}\mathrm{ar}(\bar{X}_L) = \Omega_L$
was numerically estimated by the sample variance of

$$\left\{\bar{X}_L^{(1)}, \bar{X}_L^{(2)}, \ldots, \bar{X}_L^{(B)}\right\},$$

the quantity $\mathbb{E}(S_L^2/n_L) = \tilde{\Omega}_L$ was numerically estimated by the sample mean
of

$$\left\{\left(\frac{S_L^{(1)}}{\sqrt{n_L}}\right)^2, \left(\frac{S_L^{(2)}}{\sqrt{n_L}}\right)^2, \ldots, \left(\frac{S_L^{(B)}}{\sqrt{n_L}}\right)^2\right\},$$

and the probability π_L was numerically estimated by the percentage of intervals $I^{(1)}, I^{(2)}, \ldots, I^{(B)}$ which missed the true value 0. Finally, these numeric estimates were superimposed onto the theoretical plots of Ω_L, $\widetilde{\Omega}_L$, and π_L; see Figure 8.3.   □

Appendix 8.A Intuition of Benjamini-Hockberg

This appendix aims to provide an intuitive understanding of the Benjamini-Hockberg testing procedure (Fun Box 4; section 8.2.1).

Just as the frequentist approach does not treat the model parameter θ as a random variable, even though we do not know its true value, so classic significance testing does not treat the (true-or-false) status of the null hypothesis H_0 as random, even though we do not know what it is. But what if we *do* treat it as random? Then H_0 is not just either true or false; it will have a certain probability of being true! This is a fundamental conceptual shift.

For any H_0, let U be the p-value associated with testing it. Recall that we reject H_0 if the p-value is small (see section 8.1.2), say, if $U \le u$. We can now apply Bayes law to conclude

$$\mathbb{P}(H_0|U \le u) = \frac{\overbrace{\mathbb{P}(U \le u|H_0)}^{\stackrel{(\dagger)}{=}u}\overbrace{\mathbb{P}(H_0)}^{\le 1}}{\mathbb{P}(U \le u)}. \tag{8.7}$$

Here, $\mathbb{P}(H_0)$ can be thought of as the *prior* probability of H_0 being true and $\mathbb{P}(H_0|U \le u)$ the *posterior* probability of it being true, *given that it has been*

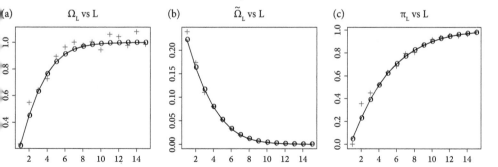

Figure 8.3 Points coded by "+" are numeric estimates based on 500 simulations to "verify" the correctness of these results.

Source: authors.

rejected. It is very important to note that this posterior probability is *not* the same as the classic notion of the significance level. (If this distinction is not clear to you, stop reading and think about it before continuing.)

The equality marked by (†) above is due to the conclusion from Exercise 8.3 (section 8.1.2). (Do the exercise and think it through.) If we evaluate (8.7) at $u = u_{(i)}$, we get

$$\mathbb{P}(H_0 | U \leq u_{(i)}) \leq \frac{u_{(i)}}{\mathbb{P}(U \leq u_{(i)})}. \tag{8.8}$$

However (and this is another key insight), with a total of N null hypotheses being tested together, we observe not just one but multiple realizations of U, i.e., $u_1, u_2, \ldots, u_N$, and the (otherwise unknown) denominator can now be estimated by

$$\widehat{\mathbb{P}}(U \leq u_{(i)}) = \frac{1}{N} \sum_{j=1}^{N} I(u_j \leq u_{(i)}) = \frac{i}{N},$$

where $I(\cdot)$ is a binary indicator function. This is because $u_{(i)}$ is, by definition, the i-th smallest of all $u_1, u_2, \ldots, u_N$.

Plugging this back into (8.8), we see that, if $u_{(i)} \leq i\alpha/N$, then an estimated version of (8.7) will satisfy

$$\widehat{\mathbb{P}}(H_0 | U \leq u_{(i)}) \leq \frac{u_{(i)}}{i/N} \leq \alpha.$$

Intuitively, therefore, the Benjamini-Hockberg procedure can be regarded as trying to reject, sequentially but greedily, as many null hypotheses as possible, *provided that* the posterior probability of the last rejected hypothesis being true—conditional on it being rejected—can still be estimated to be no larger than α.

PART IV
APPENDIX

Some Further Topics

We started this book by laying down, as early as Chapter 1, a specific framework for approaching data-analytic tasks: first, choose a probability model to describe the data-generating process; second, use the data themselves to estimate the parameters (and other unobservable quantities) in the model; third, rely on the estimated model to describe what has been learned from the data, including how predictions can be made. For much of the book, we then essentially proceeded with the assumption that not only do we already have some data to start with but also that the probability model for describing the data-generating process has been chosen for us. Neither of these assumptions hold up in practice. In this Appendix, we briefly outline some further topics that address these issues.

A.1 Graphical models

Suppose $X_1, \ldots, X_d$ are a bunch of random variables that we'd like to describe with a probability model. One way to come up with such a model is by postulating—often recursively—that certain subsets of variables (e.g. X_α and X_β) are *conditionally independent* given others (e.g. X_γ), where $\alpha, \beta, \gamma \subset \{1, 2, \ldots, d\}$ are non-overlapping subsets.

Example A.1. For example, after completing a project successfully, members of a five-person team will each receive a certain amount of bonus, say, $X_1, \ldots, X_5$. How can we model their joint distribution, $f(x_1, \ldots, x_5)$?

For convenience, let's simply refer to the team members as A1, B2, C3, D4, and E5. The following knowledge of the team structure may provide us with some hints: A1 is the team leader; B2, C3, and D4 all work directly under A1 but not with each other; E5 is an apprentice supervised by B2. Figure A.1 shows a graphical representation of this structure. Based on such information, we may postulate

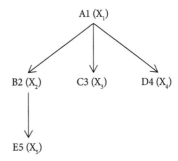

Figure A.1 A five-person team and how it works.
Source: authors.

$$f(x_1,\ldots,x_5)$$

$$\overset{(a)}{=}\quad f(x_1)\times f(x_2,x_3,x_4,x_5|x_1)$$

$$\overset{(b)}{=}\quad f(x_1)\times [f(x_2,x_5|x_1)\times f(x_3|x_1)\times f(x_4|x_1)]$$

$$\overset{(c)}{=}\quad f(x_1)\times [f(x_2|x_1)\times f(x_5|x_2,x_1)]\times f(x_3|x_1)\times f(x_4|x_1)$$

$$\overset{(d)}{=}\quad f(x_1)\times f(x_2|x_1)\times f(x_5|x_2)\times f(x_3|x_1)\times f(x_4|x_1). \qquad (A.1)$$

Step (a) follows from the definition of the conditional distribution. Step (b) encodes our assumption that (X_2,X_5), X_3, and X_4 are conditionally independent, given X_1. Step (c) follows, again, from the definition of the conditional distribution. (Convince yourself.) Step (d) encodes our assumption that, given X_2, X_5 is conditionally independent of X_1.

Notice that the conditional independence assumptions encoded in steps (b) and (d) are quite natural in this context, and they greatly simplify the joint distribution, $f(x_1,\ldots,x_5)$. It is now much easier to postulate models for each component. For example, we may choose to model them as

$$X_1\sim N(\mu_1,\sigma_1^2),\quad X_2|X_1\sim N(\mu_2+\theta_2 X_1,\sigma_2^2),\quad X_5|X_2\sim N(\mu_5+\theta_5 X_2,\sigma_5^2),$$
$$X_3|X_1\sim N(\mu_3+\theta_3 X_1,\sigma_3^2),\quad X_4|X_1\sim N(\mu_4+\theta_4 X_1,\sigma_4^2).$$

[*Think: How can we interpret the parameters $\mu_1,\ldots,\mu_5$, $\theta_2,\ldots,\theta_5$, $\sigma_1^2,\ldots,\sigma_5^2$? What does it mean if $\hat\theta_3 > \hat\theta_4$ and $\hat\sigma_3^2 < \hat\sigma_4^2$?*]   □

It is by no means a coincidence in Example A.1 that the joint distribution turns out to have a factorization (A.1) that matches exactly the flow of arrows in Figure A.1; it is an example of a *graphical model*.

A.2 Regression models

Now, suppose that, among a bunch of random variables, we are particularly interested in being able to predict one of them (say, Y) by using others (say, $X_1, \ldots, X_d$) as predictors. In this context, it is very common to work simply with a *conditional* probability model for $Y|X_1,\ldots,X_d$, rather than a joint model for $(Y,X_1,\ldots,X_d)$.

Typically, Y is assumed to follow some probability distribution, say, $f(y;\theta,\phi)$, with a certain "key" parameter θ that is a function $h(x_1,\ldots,x_d)$ of the predictors and some "other" parameter ϕ that is not. (A scenario of this type was presented earlier in Example 5.2.) Of course, we do not presume to know the function h, and it must be estimated from data. Let $\hat h$ denote the corresponding estimate; and let $\boldsymbol{x}\equiv(x_1,\ldots,x_d)^\top$.

Example A.2. If $Y\in\mathbb{R}$, it is customary to postulate that it follows a $N(\mu,\sigma^2)$ distribution. The "key" parameter is usually taken to be μ and the "other" parameter σ^2, that is,

$$\theta\equiv\mu=h(\boldsymbol{x})\quad\text{and}\quad\phi\equiv\sigma^2.$$

Then,

$$\mathbb{E}(Y)=\mu=h(\boldsymbol{x}).$$

Given any x_{new} and having estimated h from data, we may thus predict the corresponding Y_{new} by $\widehat{Y}_{new} = \widehat{h}(x_{new})$.   □

Example A.3. If $Y \in \{0, 1\}$, it is customary to postulate that it follows a Bernoulli(p) distribution. The "key" parameter is usually taken to be

$$\theta \equiv \log \frac{p}{1 - p} = h(x),$$

and there is no "other" parameter ϕ remaining in the model. (The parameter θ above is called the *logit transform* of p; it is actually the more "natural" parameter for the Bernoulli distribution. Some reasons behind this have to do with the so-called *exponential family* of probability distributions, of which the Bernoulli distribution is a member, but we won't go into these details here.) Then,

$$\mathbb{P}(Y = 1) = p = \frac{e^{h(x)}}{1 + e^{h(x)}} \geq 1/2 \quad \text{if, and only if,} \quad h(x) \geq 0.$$

Given any x_{new} and having estimated h from data, we may thus predict $\widehat{Y}_{new} = 1$ if $\widehat{h}(x_{new}) \geq 0$, and $\widehat{Y}_{new} = 0$ if $\widehat{h}(x_{new}) < 0$.   □

   In both Examples A.2 and A.3 above, it is clear that the main modeling effort will be concentrated on the function $h(x)$ itself. Usually, a "beginner" choice of $h(\cdot)$ is a *linear model*,

$$h(x) = \beta_0 + \beta_1 x_1 + \ldots + \beta_d x_d.$$

This may seem easy enough, but there are already difficult statistical issues when d is very large. A slightly more general choice of $h(\cdot)$ may be an *additive model*,

$$h(x) = \beta_0 + h_1(x_1) + \ldots + h_d(x_d),$$

where each $h_j(\cdot)$ is univariate function with a certain smoothness requirement, for example, $\int [h_j''(t)]^2 dt < s_j$ for some fixed $s_j > 0$ (see Remark 5.5; section 5.1.2). An even more flexible choice of $h(\cdot)$ may be a multilayer *neural network*, which is currently very popular. One can also take a Bayesian approach by treating the unknown function h as a random object and putting a prior distribution on it.

A.3 Data collection

Increasingly, the main job of data scientists is to make sense of data that are already in front of them, but that's not how statistics was practiced traditionally. A classic statistical question concerns how we should collect the data in the first place because, if this were not carried out properly, the analysis afterwards would become futile.

   For example, if we wanted to have an idea of how much exercise people are getting nowadays, and we simply conducted a survey by asking every customer at a local coffee shop on a Sunday morning, then, even if our survey went very well, we would still obtain only a partial answer at best—namely, how much exercise *coffee drinkers*, not the general public, are getting. (In fact, such a survey will most likely miss the most avid of exercisers, too since, on a Sunday morning, they'd probably be working out at a gym rather than showing up at a coffee shop!)

This example elucidates what "futile" means in our context—our work would become futile if the data we collected did not allow us to properly answer the original question that prompted us to collect the data in the first place. Therefore, the inquiry about how to collect data must necessarily be directed by what question we want the data to help us answer; the inquiry itself would become completely vacuous otherwise.

Example A.4.   For example, let's look at survey sampling in a little more detail. Suppose there is a finite *population* (of fixed values), $\{x_1, \ldots, x_N\}$, and we simply want to know what their average,

$$\theta \equiv \frac{1}{N} \sum_{i=1}^{N} x_i \tag{A.2}$$

is equal to. We cannot afford to query every value; for example, N may be too big, so we take a *sample* (e.g. $\{x_3, x_{15}, \ldots, x_{523}\}$), in which case, we shall denote the corresponding index set $\{3, 15, \ldots, 523\}$ by S. We simply estimate the unknown quantity θ by its "sample analogue", $\hat{\theta} = (1/n) \sum_{i \in S} x_i$, where $n = |S|$ is the sample size. Clearly, the randomness of $\hat{\theta}$ here is due to the uncertainty of which values $x_1, \ldots, x_N$ will end up in our sample S, not the values $x_1, \ldots, x_N$ themselves.

The key for analyzing the statistical behavior of $\hat{\theta}$ is to define an indicator random variable S_i for every $i = 1, \ldots, N$, and to express $\hat{\theta}$ as

$$\hat{\theta} = \frac{1}{n} \sum_{i=1}^{N} x_i S_i, \quad \text{where} \quad S_i = \begin{cases} 1, & i \in S; \\ 0, & i \notin S. \end{cases} \tag{A.3}$$

This is a linear function of $S_1, \ldots, S_N$, and we can compute its expectation as

$$\mathbb{E}(\hat{\theta}) = \frac{1}{n} \sum_{i=1}^{N} x_i \underbrace{\mathbb{E}(S_i)}_{\mathbb{P}(i \in S)} = \frac{1}{n} \sum_{i=1}^{N} x_i \underbrace{\left[\frac{n}{N}\right]}_{\mathbb{P}(i \in S)} = \frac{1}{N} \sum_{i=1}^{N} x_i = \theta,$$

which shows the sample analogue is an unbiased estimator—*provided that* we can really ensure $\mathbb{P}(i \in S) = n/N$ for every $i = 1, \ldots, N$ when we take our sample (or conduct our survey). But what happens if this is not the case?

Suppose the population can be partitioned into two groups,

$$\{ \underbrace{x_1, \ldots, x_A}_{\text{Group 1}}, \underbrace{x_{A+1}, \ldots, x_N}_{\text{Group 2}} \},$$

and the probability $\mathbb{P}(i \in S)$ differs depending on whether i belongs to the first group ($i \le A$) or the second one ($i > A$). We can rewrite

$$\theta = \frac{1}{N} \left[\sum_{i=1}^{A} x_i + \sum_{i=A+1}^{N} x_i \right] \tag{A.4}$$

to reflect the group structure explicitly and, for each i, define two new (conditional) indicators, S_i^- and S_i^+, respectively, for $i \leq A$ and $i > A$, such that

$$\mathbb{P}(S_i^- = 1) \equiv \mathbb{P}(i \in S | i \leq A) \quad \text{and} \quad \mathbb{P}(S_i^+ = 1) \equiv \mathbb{P}(i \in S | i > A).$$

This allows us to express $\widehat{\theta}$ as

$$\widehat{\theta} = \frac{1}{n} \left[\sum_{i=1}^{A} x_i S_i^- + \sum_{i=A+1}^{N} x_i S_i^+ \right]. \tag{A.5}$$

However, by the law of total probability, we must have

$$\underbrace{\mathbb{P}(i \in S)}_{n/N} = \underbrace{\mathbb{P}(i \in S | i \leq A)}_{p} \underbrace{\mathbb{P}(i \leq A)}_{A/N} + \underbrace{\mathbb{P}(i \in S | i > A)}_{q} \underbrace{\mathbb{P}(i > A)}_{(N-A)/N}. \tag{A.6}$$

One can easily verify from Equation (A.6) that, if $q = p$, this simply implies $p = q = n/N$, which guarantees $\widehat{\theta}$ to be unbiased, as we have shown above.

It is interesting to see what happens now if

$$q = kp \quad \text{for some fixed } k > 1 \quad \text{and} \quad x_i = \begin{cases} 1, & i \leq A; \\ 0, & i > A. \end{cases}$$

Equation (A.4) now shows that the quantity we are interested in, $\theta = A/N$, is merely the proportion of the population belonging to the first group; and Equation (A.5) now implies that

$$\mathbb{E}(\widehat{\theta}) = \frac{1}{n} \left[\sum_{i=1}^{A} x_i \underbrace{\mathbb{E}(S_i^-)}_{p} + \sum_{i=A+1}^{N} x_i \underbrace{\mathbb{E}(S_i^+)}_{q} \right] = \frac{Ap}{n}. \tag{A.7}$$

But if $q = kp$, we see that

$$p = \frac{n}{kN - (k-1)A}$$

by Equation (A.6), and plugging this into Equation (A.7) yields

$$\mathbb{E}(\widehat{\theta}) = \frac{\theta}{k - (k-1)\theta} \neq \theta. \tag{A.8}$$

Hence, if 53% of Americans actually supported Donald Trump in the 2016 US presidential election ($\theta = 0.53$), but those who did not support him were, for whatever reason, twice as likely to end up in opinion polls than those who did ($k = 2$), then, on average the polls would estimate that only $(0.53)/[2 - (2 - 1)(0.53)] \approx 0.36$ or 36% of Americans supported him—a downward bias of a whopping 17 percentage points! Although the numeric values here are hypothetical, qualitatively speaking this is what really happened in 2016 and why the polls were so wrong that year. □

Example A.5.   Another ubiquitous scientific question is whether A causes B, for example, A = smoking and B = cancer. To answer such a question, it is *not* enough to simply look at people who smoke and those who don't because there could be other significant differences between these two groups of people that are actually responsible for their cancer (or the lack thereof), instead of whether they smoke or not. This is a very important point.

For example, perhaps smokers are just more easily stressed, and that's why they became smokers in the first place. If so, then it's not clear if their higher rate of cancer is really caused by smoking or merely by the frequently elevated levels of stress hormones in their bodies.

There are many such factors that can make it equally difficult for us to answer the question of whether smoking causes cancer; these are known as *confounding factors*. How can we go about collecting data in such a way that diminishes their impact? What does it mean, and why is it important, to *randomize*? How can we do that most effectively? What can we do if it's impossible to randomize in practice? These are questions of *experimental design*, another important area of statistics that we haven't covered in this book.   □

Perhaps surprisingly, these seemingly classic topics of statistics are experiencing a renewal in the contemporary age. For example, instead of telephone surveys, we can now conduct surveys on social media. By encouraging people to invite their friends to participate, we can reach a potentially large audience with relatively little cost, but surely this kind of data collection process will entail a different kind of bias altogether? As another example, unlike television or newspaper advertisements, online advertisements are easily randomizable. A company website can be programmed so that, when an online customer comes along, it will display a randomly selected advertisement from its collection; and if the collection is carefully designed, the company may be able to reliably conclude, for instance, that certain advertising elements are more effective in the morning while others are more effective after lunch hour!

Epilogue

The essence of this text—an extremely fast-paced introduction to statistics—necessarily means that most of the materials are fairly standard. The only novel aspect and contribution of this book lie in what materials are selected, as well as how they are organized and presented, so as to facilitate most effectively a crash course of this nature. These efforts have been explained and summarized in the Prologue section.

The opening and closing parts of the book, however, are emphatically personal. The very first set of examples used in the text, Examples 1.1, 1.2, and 1.3, are based on my own joint work [4] with a former PhD student and a collaborator. The very last section of the main text (section 8.2.2) and formula (A.8) in Appendix A.3 are also based entirely upon my own idiosyncratic thoughts.

I owe Example 5.5 to Professors Jerry Lawless and Steve Brown. I first encountered it from a set of teaching notes that they had created. I liked it so much that I immediately started using it in my own classes. Since then, I have expanded it into a running series which, apart from their original example, now also encompasses Example 5.2, Example 5.4, Exercise 5.6, and Example 5.6.

Creating a crash course like this has been an enormous professional challenge, to say the least, but it has also given me a rare opportunity to think about the subject of statistics in broad strokes and from a unique angle. It definitely has sharpened my own understanding of what statistics is about.

I couldn't have possibly written this book without the positive influences of: Professor Jun Liu, who enthusiastically introduced me to the field of statistics; Professor David Laibson, who subconsciously inspired me to pursue a career in academia; Professors Trevor Hastie, Jerry Friedman, and Rob Tibshirani, who patiently guided me through my doctoral studies; Professors David Matthews and Mary Thompson, who generously offered me an academic position afterwards; Professors Hugh Chipman and Will Welch, who kindly mentored me during my first few years on the job; Professors Changbao Wu, Annie Qu, and Nancy Reid, who tirelessly helped to promote my career; and finally, Professor David Donoho, who earnestly encouraged me to keep on writing. I am forever grateful to these marvelous people. While I will never be able to match *any* of them in this lifetime, they will always be my role models—not only as great scholars but also as wonderful human beings.

On a personal level, I am much indebted to Vince Canzoneri, Harold Otto, Helen Tombropoulos, and Tova Wein, who have all helped me in more important ways than they perhaps realize themselves.

Finally, I would like to thank my colleagues, Professors Pengfei Li and Kun Liang, and my student, Aditya Degala, for various discussions about the manuscript, plus the wonderful team at Oxford University Press—especially Katherine Ward for her sustained enthusiasm and faith in this project during the most difficult months of the coronavirus pandemic, as well as Lilith Dorko, Hayley Miller, John Smallman, and Dan Taber, for their enduring patience and unwavering support along the way.

Now that I am writing the Epilogue, it is my hope that this text will allow many students to find learning statistics manageable and perhaps even enjoyable.

Bibliography

[1] Holland P. W., Laskey K. B., Leinhardt S. (1983), "Stochastic blockmodels: First steps", *Social Networks*, **5**(2):109–137.

[2] Snijders T. A. B., Nowicki K. (1997), "Estimation and prediction for stochastic block-models for graphs with latent block structure", *Journal of Classification*, **14**:75–100.

[3] Karrer B., Newman M. E. J. (2011), "Stochastic blockmodels and community structure in networks", *Physical Review E*, **83**:016107.

[4] Xin L., Zhu M., Chipman H. A. (2017), "A continuous-time stochastic block model for basketball networks", *Annals of Applied Statistics*, **11**:553–597.

[5] Ross S. (2007), *Introduction to Probability Models* (9th edn), Academic Press.

[6] Dauphin Y. N., Pascanu R., Gulcehre C., Cho K., Ganguli S., Bengio Y. (2014), "Identifying and attacking the saddle point problem in high-dimensional non-convex optimization", *Advances in Neural Information Processing Systems*, **27**:2933–2941.

[7] Galton F. (1886), "Regression towards mediocrity in hereditary stature", *Journal of the Anthropological Institute of Great Britain and Ireland*, **15**:246–263.

[8] Lange K. (1999), *Numerical Analysis for Statasticians*, Springer.

[9] Fraley C., Raftery A. E. (2002), "Model-based clustering, discriminant analysis, and density estimation", *Journal of the American Statistical Association*, **97**:611–631.

[10] Blei D. M., Ng A. Y., Jordon M. I. (2003), "Latent Dirichlet allocation", *Journal of Machine Learning Research*, **3**:993–1022.

[11] Efron B., Morris C. (1977), "Stein's paradox in statistics", *Scientific American*, **236**(5):119–127.

[12] Givens G. H., Hoeting J. A. (2005), *Computational Statistics*, Wiley.

[13] Roberts G. O., Rosenthal J. S. (2001), "Optimal scaling for various Metropolis-Hastings algorithms", *Statistical Science*, **16**:351–367.

[14] Young S. S., Karr A. (2011), "Deming, data and observational studies: A process out of control and needing fixing", *Significance*, **8**:116–120.

Index